RESTORING & FLYING A
SPORT PLANE ON A BUDGET

Other TAB Books by the author:

2329 *The Flight School Handbook*

RESTORING & FLYING A
SPORT PLANE ON A BUDGET

BY RANDALL BRINK

MODERN AVIATION SERIES

TAB BOOKS Inc.

BLUE RIDGE SUMMIT, PA. 17214

FIRST EDITION

FIRST PRINTING

Copyright © 1982 by TAB BOOKS Inc.

Printed in the United States of America

Reproduction or publication of the content in any manner, without express permission of the publisher, is prohibited. No liability is assumed with respect to the use of the information herein.

Library of Congress Cataloging in Publication Data

Brink, Randall.
    Restoring & flying a sport plane on a budget.

    Includes index.
    1. Airplanes—Restoration. I. Title. II. Title: Restoring and flying a sport plane on a budget.
TL671.9.B74   629.133'343   81-18348
ISBN 0-8306-2319-1 (pbk.)   AACR2

Contents

	Introduction	vii
1	**Economics of Sport Airplane Ownership**	1
	Initial Costs—Operating Costs—Maintenance Costs—Rebuilding Costs—Per Hour Costs	
2	**Choosing Your Airplanes**	11
	Taylorcraft BC 12D—Stinson 108—Piper Tri-Pacer—Piper J-3 Cub—Luscombe Silvaire—Aeronca Champion—Ercoupe / Aircoupe—Piper Colt—Cessna 120/140—All-Metal Airplanes	
3	**Lightplane Engines**	31
	Engine Logs—Top Overhaul—Major Overhaul	
4	**The Pre-Purchase Checklist**	36
	Cockpit—Wings—Engine—Fuselage—Tail—Flight Test	
5	**Long-Distance Buying**	43
6	**Getting Started**	50
	Practical Choice of Fabric Covering—Work Area Considerations—Disassembly—Inspecting the Airframe—Preparing Wings and Control Surfaces—Preparation for Fuselage Cover	
7	**Applying the Fabric**	109
	Ceconite Process-General—Repairs—Annual Inspection—Ceconite Appendix A: Standard Finish—Ceconite Appendix B: Alternative Finish—Ceconite Appendix C: Alternative Finish—Helpful Hints—Advantages of 7600 Finish—Reassembling the Aircraft—In the Air Again	
8	**Refurbishing Metal Aircraft**	137
	Aluminum Surface Painting—Steel Surface Painting—Magnesium Surface Painting—Treating Oxidized or Corroded Aluminum—Inaccessible Areas and Corrosion Protection—Painting Interior Surfaces—Finishing Fiberglass Components—Refinishing Old Paint—Primers and Finish Coats—Spraying Equipment—Miscellaneous Information	

9 Accessories — 147
Floatplane Flying—Winter Flying on Skis—Electronics—Ultralights: A Low Cost Alternative

10 Other Considerations — 159
Rebuilding From Salvage—Engine Swapping—Learning to Fly in Your Own Plane—Resale Value—Maintenance: What You Can Do—Summary

Appendix: State Aviation Departments — 169

Sources of Materials and Supplies — 175

Index — 181

Introduction

Once smitten with the flying bug, it's a rare individual who does not put flying before all other recreational activities. Weekends, holidays, and any time not spent at gainful employment is likely to be spent at the airport, if not in an airplane. Once you have learned to fly, however, it won't take long for you to realize that flying is extremely expensive, especially if it means having to rent an airplane. The instant this realization enters your mind, the seed of desire to own an airplane begins to germinate. Sooner or later, this tiny flame of desire will blossom into a conflagration of *need* to have an airplane, available to fly at all times, upon your whim.

Now, of course there are many different reasons or rationalizations for owning an airplane. Most of them deal with the utility and convenience of an airplane as a transportation machine. If you have a need for fast, efficient long-distance travel, there is no better vehicle than an airplane. But if you cannot justify owning an airplane for business or personal transportation, or simply want to fly for the pure and simple pleasure of being aloft, and if you can not possibly afford to buy a new, or even fairly new, dream machine, there is still hope. The door to fulfilling your dream of ownership is not eternally closed. You do not have to be a corporate vice president or own oil wells to partake of the thrill of flying. There are conventional aircraft available for less than we are currently accustomed to shelling out for a late-model automobile. Also, there are available some new cost-conscious alternatives to conventional aircraft that are discussed later. For now, however, there are perfectly flyable conventional airplanes which can be purchased for between $4000 and $6000. They may be old and they may need some or a lot of TLC (tender loving care), but for sport flying on a budget they are unbeatable.

Chapter 1
Economics of Sport Airplane Ownership

Let's take a realistic look at the economics of owning and operating an airplane and then answer the big question: *Is sport flying affordable?*

When contemplating becoming an aircraft owner, try to put aside your enthusiasm for buying an aircraft long enough to consider carefully the costs of owning an airplane.

Initial Costs

First, there is the initial price of the airplane itself. Then, there are other, frequently overlooked costs at the time of purchase. You may incur costs in searching for and bringing home your new-found treasure. (I discuss "long-distance buying" in a later chapter.) In addition, there are such costs as sales tax, registration fees, and title search fees.

If you live in a state that levies a sales tax, you will have to pay that tax on the purchase price of the aircraft. The percentage rate of sales tax charged varies from state to state, of course, but could run to $200 or $300 on an aircraft sold in the $4000 to $6000 range.

Registration varies also from state to state, but it is usually a small fee, such as $25 for a light single-engine airplane. It is a good idea to ask the aviation division of your state government about registration of your airplane. (For a list of the aviation departments in each state, see the Appendix.)

Finally, on the subject of a title search, let me point out that there is no law *requiring* a prospective aircraft buyer to run a title search on an airplane, but frankly, anyone who buys an airplane without one is courting disaster. Bear in mind that when dealing with aircraft, you will not receive a printed title such as is the case with an automobile. The aircraft title, along with any record of liens, attachments, or joint ownership remains with the FAA in Oklahoma City (Figs. 1-1, 1-2). Therefore, there is no way that you could immediately detect any encumbrance on the aircraft, which may have been placed there by someone to whom a previous owner owed money, or whether there was a joint owner of the aircraft. Liens must be registered with the FAA Aircraft Registration Branch in Oklahoma City, Oklahoma, to be perfected. However, the only way that a buyer would know about a lien is to have a thorough title search done. it is an easy procedure to accomplish. Either call a title search firm that specializes in FAA aircraft searches and advertises in various aviation magazines and newspapers, or contact the Aircraft Owners and Pilot's Association (AOPA):

>AOPA
>Box 19244 Southwest Station
>Oklahoma City, OK 73144
>Telephone: (409) 682-2511 Telex: 747-265

The AOPA organization is one of the strongest owner/pilot groups in the nation and has an office adjacent to the FAA headquarters in Oklahoma City. For a $15 to $20 fee (charged to members and non-members respectively), they will thoroughly check the file on the airplane you plan to purchase and report back to you. Usually within a day they can notify you of any liens or other encumbrances on the title of the airplane you requested searched. Once this search has been completed, you can rest reasonably assured that no one will show up after the sale with a claim against the aircraft.

Operating Costs

Some operating costs, like sales tax and registration fees, also vary depending on the area of the country in which you live.

Fuel, for example, is higher in the northeastern U.S. than in the south; 80/87 octane aviation fuel, the kind used in most light single-engine aircraft, varies from $1.48 to $2.00 per gallon. In some areas, this grade of avgas is not even available, in which case, the higher-priced 100/130 avgas may have to be used. But when considering the fuel factor, the older airplanes are still very attractive, since most small two seaters only use about four to six gallons of fuel per hour.

Another geographical variance in operating costs are those of parking or hangaring an airplane. Some airports charge nothing to keep an airplane out on the grass tiedowns or even on the paved tiedown line (Fig. 1-3). At other fields, especially at busy metropolitan airports, the tiedown fee may run as high as $75 per month. Hangar space can cost from about $30 per month to as high as $400

*U.S. GOVERNMENT PRINTING OFFICE: 1977-774-297

FORM APPROVED: OMB NO. 04-R0076

UNITED STATES OF AMERICA
DEPARTMENT OF TRANSPORTATION - FEDERAL AVIATION ADMINISTRATION

AIRCRAFT BILL OF SALE

FOR AND IN CONSIDERATION OF $ THE UNDERSIGNED OWNER(S) OF THE FULL LEGAL AND BENEFICIAL TITLE OF THE AIRCRAFT DESCRIBED AS FOLLOWS:

UNITED STATES REGISTRATION NUMBER N

AIRCRAFT MANUFACTURER & MODEL

AIRCRAFT SERIAL No.

DOES THIS DAY OF 19
HEREBY SELL, GRANT, TRANSFER AND DELIVER ALL RIGHTS, TITLE, AND INTERESTS IN AND TO SUCH AIRCRAFT UNTO:

Do Not Write In This Block
FOR FAA USE ONLY

PURCHASER

NAME AND ADDRESS
(IF INDIVIDUAL(S), GIVE LAST NAME, FIRST NAME, AND MIDDLE INITIAL.)

DEALER CERTIFICATE NUMBER

AND TO EXECUTORS, ADMINISTRATORS, AND ASSIGNS TO HAVE AND TO HOLD SINGULARLY THE SAID AIRCRAFT FOREVER, AND WARRANTS THE TITLE THEREOF.

IN TESTIMONY WHEREOF HAVE SET HAND AND SEAL THIS DAY OF 19

SELLER

NAME (S) OF SELLER (TYPED OR PRINTED)	SIGNATURE (S) (IN INK) (IF EXECUTED FOR CO-OWNERSHIP, ALL MUST SIGN.)	TITLE (TYPED OR PRINTED)

ACKNOWLEDGMENT (NOT REQUIRED FOR PURPOSES OF FAA RECORDING: HOWEVER, MAY BE REQUIRED BY LOCAL LAW FOR VALIDITY OF THE INSTRUMENT.)

ORIGINAL: TO FAA

AC FORM 8050-2 (8-76) (0052-629-0002)

Fig. 1-1. You must have the bill of sale filled out by the seller and the original will be sent to the FAA in Oklahoma City, OK. This constitutes the transfer of ownership of the airplane.

per month. The actual cost at your airport will help you decide whether or not to hangar your airplane. But remember this: airplanes that are fabric covered will not last as long outdoors as indoors. Heat, cold, sunlight, ice, and air pollution will constantly attack the fabric, no matter how good the cover. Hangaring *vs* keeping the airplane outside could make a difference of several years in the useful life of your fabric. Also, the general appearance of your airplane is easier to maintain if the airplane is kept sheltered from the elements. It is important, therefore, that you weigh the costs and the effects of climate in your area and decide accordingly, whether or not it is economically more wise to tiedown outside or cough up the additional bucks for a hangar. Sun is fabric's worst enemy.

Then there is the insurance cost, which not only varies according to the area in which you are flying, but also varies according to your flying experience. As the experience of the pilot or pilots increases, the cost of insurance decreases. Insurance costs are a major consideration; and, two points should be made here: (1) Liability insurance is an absolute necessity. No one, no matter how wealthy, can afford to pay a huge judgment in the event of an accident where personal injury or property damage is involved. (2) Hull coverage, on the other hand, is up to the individual discretion of the owner. There is no requirement to carry hull insurance on an airplane. This coverage is similar to the comprehensive insurance that you carry on your automobile to provide for its replacement or repair in the event of an accident. If you're willing to withstand the loss of the aircraft, you may elect not to have hull coverage. Check with a well-known aviation insurance underwriter for your insurance needs. (See TAB Book 2316, *Pilot's Guide to Aviation Insurance*, for complete details.)

Maintenance Costs

Probably the most crucial operating cost involved in evaluating the economics of aircraft ownership is the cost of maintenance.

Aircraft maintenance is done in ways completely different than anything you are likely to have experienced before, and so involves considerations you might easily overlook in deciding whether ownership is for you. The main difference between aircraft maintenance and automotive or other maintenance procedures you may be familiar with is the concept of preventive maintenance involving routine inspections, which are required by law and which are somewhat more expensive than you might at first imagine.

☆ U.S. GOVERNMENT PRINTING OFFICE 1977 — 774-155

FORM APPROVED OMB NO. 04-R0076

UNITED STATES OF AMERICA
DEPARTMENT OF TRANSPORTATION - FEDERAL AVIATION ADMINISTRATION
AIRCRAFT REGISTRATION APPLICATION

CERT. ISSUE DATE

UNITED STATES REGISTRATION NUMBER **N**

AIRCRAFT MANUFACTURER & MODEL

AIRCRAFT SERIAL No.

FOR FAA USE ONLY

TYPE OF REGISTRATION (Check one box)

☐ 1. Individual ☐ 2. Partnership ☐ 3. Corporation ☐ 4. Co-Owner ☐ 5. Gov't.

NAME OF APPLICANT (Person(s) shown on evidence of ownership. If individual, give last name, first name, and middle initial.)

ADDRESS (Permanent mailing address for first applicant listed.)

Number and street: _____
Rural Route: _____   P. O. Box: _____

CITY	STATE	ZIP CODE

☐ CHECK HERE IF YOU ARE ONLY REPORTING A CHANGE OF ADDRESS

ATTENTION! Read the following statement before signing this application.

A false or dishonest answer to any question in this application may be grounds for punishment by fine and/or imprisonment (U.S. Code, Title 18, Sec. 1001).

CERTIFICATION

I/WE CERTIFY that the above described aircraft (1) is owned by the undersigned applicant(s), who is/are citizen(s) of the United States as defined in Sec. 101(13) of the Federal Aviation Act of 1958; (2) is not registered under the laws of any foreign country; and (3) legal evidence of ownership is attached or has been filed with the Federal Aviation Administration.

NOTE: If executed for co-ownership all applicants must sign. Use reverse side if necessary.

EACH PART OF THIS APPLICATION MUST BE SIGNED IN INK.

SIGNATURE	TITLE	DATE
SIGNATURE	TITLE	DATE
SIGNATURE	TITLE	DATE

NOTE: Pending receipt of the Certificate of Aircraft Registration, the aircraft may be operated for a period not in excess of 90 days, during which time the PINK copy of this application must be carried in the aircraft.

AC FORM 8050-1 (8-76) (0052-00-628-9004)

Fig. 1-2. The aircraft registration application is filled out and sent to Oklahoma City. The pink copy becomes your temporary aircraft registration, which must be carried in the airplane. This is the proof of ownership document, as there is no title, as such, for an airplane.

Fig. 1-3. A tiedown will cost from $10 up at most general aviation airports. Your airplane is better kept in a hangar, but these can run from $30 to over $300 in some places.

Also, unlike what you're accustomed to in automotive maintenance, the FAA occasionally issues "Airworthiness Directives," commonly referred to as "ADs," on an airplane when a particular need for repair or inspection of a component part is deemed necessary. These ADs are *mandatory*. The problem you can anticipate with older airplanes is that some of the ADs may not have been complied with, yet, you as a buyer would probably not know this unless you check with a maintenance shop and obtain a current list of the AD notes pertaining to the aircraft you are buying. Then, you can check the aircraft and engine logbooks to ascertain that all the ADs have been complied with.

The Federal Aviation Administration requires that all aircraft be inspected annually by a licensed aircraft mechanic and "signed off" by an authorized inspector. For most light aircraft, this inspection may cost between $75 and $300 per year, depending on the amount of shop work that is required to maintain the aircraft in an airworthy condition. However, you can do much of the work of maintaining your aircraft under the supervision of an FAA certified mechanic. Beware: some mechanical work *must* be done by a mechanic, or under the *direct* supervision of a mechanic. Check with your local fixed based operator or aircraft maintenance shop before attempting to do any maintenance on an aircraft for which you are not qualified. Beyond changing the oil or performing one of the other 24 FAA-approved preventive maintenance tasks, an FAA licensed mechanic is required either to do the work or closely supervise it. An "A&P" mechanic must describe any work done on the aircraft in the airframe or engine logbook, and he must sign it off, acknowledging that he is satisfied that the job was done properly within the framework of acceptable aircraft maintenance procedures. Few mechanics—and *no* reputable, professional one—would sign a logbook for a job they did not personally do or at least closely supervise.

In the case of fabric-covered aircraft, the annual inspection will include a test of the fabric for strength. Aircraft fabrics slowly deteriorate with age, and each year they must be tested to ascertain that they maintain minimum strength—56 lbs tensile strength in both directions. This test may be performed with a punch tester, which measures the amount of force required to punch a hole through the fabric and indicates the amount of strength left. Each wing is tested, as is the fuselage and each control surface. Any part failing to show up "in the green," within the range of satisfactory strength, must be recovered before further flight. The cost of having a small aircraft recovered (such as an Aeronca 7AC Champion or Piper J-3 Cub) has reached the $3500 to $5000 mark. It is of considerable consolation to note, though, that about 80 percent of the cost of having an airplane restored is labor that you could do yourself under the supervision of a qualified mechanic. Since few people who own these light, inexpensive ships can afford, to spend that kind of money, they are faced with basically two alternatives when the fabric needs to be replaced: do the job themselves under the direct supervision of an FAA licensed mechanic, or sell the airplane. This is where you come in.

Rebuilding Costs

We are back, now, to the first section of this book on "initial costs" and back to one of the basic premises of this book: when an aircraft reaches a point where it requires new fabric to meet the requirements of airworthiness, that aircraft frequently is placed on the market by an owner who cannot undertake the job himself or afford to hire it done by an FAA repair station.

You are a bargain hunter who wants to fly for the pure joy of flying. You want a basic flying machine that you can afford, one that costs but a few dollars per hour to fly, and one that is easy and fun and will not eat you up in maintenance costs. You are relatively handy with tools and do not mind hard, tedious (though rewarding) work. You don't mind getting dirty working with dope and fabric and spending a small amount of money to get the project underway. You are a sport flyer about to enter the wonderful world of aircraft ownership on a budget.

Let's take a look at the approximate material costs for rebuilding a fabric-covered aircraft. Following is a breakdown of the approximate amount and description of the materials, and their respective costs, to recover the average light two-place, single-engine airplane, such as a Piper J-3 Cub, Aeronca 7AC Champ, or Taylorcraft. These materials and approximate prices are quoted from the "Dac-Proof" system by Cooper Aviation Supply Company, of Elk Grove Village, Illinois, one of the largest suppliers of recovering materials for light aircraft (Table 1-1).

This process utilizes Ceconite 102, a lightweight synthetic fabric (Dacron) that is easy to apply, fairly inexpensive, and extremely strong and resistant to deterioration. This brand of synthetic fabric is widely used by amateur rebuilders and is given here only as an example of *one* type and brand of system. There are many others to choose from, such as Stits Polyfiber, Razorback, and others.

Bear in mind that these material estimates are only a general guideline, as are the prices estimated here which are, of course, subject to change at any time. Consult your mechanic/supervisor for exact quantities and methods (See Sources of Materials and Supplies).

Per Hour Costs

Now that we have examined the cost factors involved in owning, operating, and maintaining an aircraft, the bottom line can be seen more clearly in the following breakdown of a cost per hour estimate. This example uses the cost of operation of an Aeronca Champion:

Fuel (4.2 gallons/hour; $1.50/gallon)	$ 6.38
Oil (.2 quarts/hour; $1.50/quart)	.30
Maintenance and inspection	1.75
Engine reserve	1.75
Insurance ($500/year)	1.66
Tiedown ($25/month)	1.00
Total Hourly Estimate	$12.84

In computing these figures, an average annual utilization figure of 300 hours was used. This represents about how many hours the average sport aircraft pilot will fly an airplane of this type in an average year.

The fuel consumption rate used in the example is very typical of airplanes of this type. However, the cost per gallon of fuel is a constantly changing value. The consolation here, though, is that these airplanes use less of it by far than anything currently in production (with the exception of the amazing ultralight airplanes which will be discussed in a later chapter).

Table 1-1. Approximate Estimate of Recovery Costs for an Average Light Aircraft.

Quantity	Description	Price Each	Total
5 gal.	300 DAC Proofer	$ 19.25	$ 96.25
15 gal.	301 Spray Fill	17.35	260.25
5 gal.	131 Cab Thinner	6.75	23.75
1 gal.	132 Cab Retarder	12.00	12.00
5 gal.	Butyrate Color	15.25	76.25
1 qt.	Butyrate Color for Trim	6.95	6.95
1 gal.	1500 Clear Cab Dope	15.50	15.50
2 qt.	Super Seam Cement	7.80	15.60
1 rl.	D-693 Ceconite Rib Stitch Cord	9.80	9.80
4 rl.	2″ Ceconite Tape	8.00	32.00
2 rl.	⅜″ Ceconite Reinforcing Tape	3.20	6.40
24 ea.	Grade A Patches	.30	7.20
24 ea.	Inspection Rings	.10	2.40
1 pkg.	Drain Grommets	1.00	1.00
1 rl.	2″ Chafe Point Tape	5.65	5.65
2 sets	3″ Spray Stencils N_ _ _ _	.25	6.00
1 qt.	Zinc Chromate	13.30	13.30
1 qt.	161 Chromate Reducer	3.85	3.85
1 qt.	Synthetic Gloss Enamel	8.95	8.95
1 set	Ceconite Envelopes 102	195.16	195.16
1 set	Wood Kit (longerons, stringers, fuselage formers, floorboards)	250.00	250.00
	Miscellaneous Hardware		50.00
Total Approximate Cost of Recovery Materials			$1,108.26

As for oil, most airplanes with well-maintained engines will probably go all the way between the standard twenty-five hour oil change interval without requiring the addition of any oil. Still, it does not hurt to overestimate, rather than underestimate, costs.

The biggest variable in the example is the cost of keeping the aircraft at the local airport. Check your municipal airport for more precise costs.

The figures for maintenance and engine reserve were based, again, on the 300 hour annual utilization, and a rather large figure of $500 spent on the annual inspection and any necessary repairs during the year was used. This figure is higher than I have personally experienced with the Aeronca Champ, but an unexpected major repair might exceed the figure, so again, better safe than sorry where estimates are concerned. After all, you are trying to determine whether or not the prospect of aircraft ownership is economically feasible.

There is no hourly cost built in to cover the initial purchase of the aircraft or the cost of restoring the bird. The reason for this is that your aircraft should be considered an investment, which does not have to be amortized over the hourly utilization cost of operation. Further, the restoration costs will greatly enhance the value of your airplane over the price you paid, and, in most cases, will result in your realizing an appreciation of the aircraft. This is particularly true as the cost of fuel and new airplanes increases, making the older, well-maintained aircraft more desirable on the market. Thus, you will not only be having a ball flying for around $12 per hour (versus $20+ for a newer production aircraft), but you may actually be gaining real asset value as your airplane appreciates. What a way to invest!

So, in all probability, by reviewing these figures, applying your own local research and drawing your own (no doubt slightly prejudiced) conclusion, your answer to the hypothetical question of whether or not the sport aircraft makes economic sense for you is a resounding *yes*!

Chapter 2
Choosing Your Airplane

Now that you have definitely decided to join the ranks of airplane owners and rebuilders, comes the decision of exactly what airplane to choose. The choice of airplanes may seem endless, but a review of the goals of this venture will narrow it down considerably (Fig. 2-1).

In order to find an airplane that is inexpensive to buy and cheap to restore and operate, you are faced with a number of factors: (1) You will want an older airplane (Fig. 2-2). New airplanes are simply not available for $4000 to $6000. (2) To keep purchase, restoration, and operating costs to a minimum, you will want to choose from a few two-place airplanes and even fewer four-place airplanes that meet the requirements in terms of cost (Fig. 2-3). (3) You will be faced with a number of fabric-covered airplanes in the above mentioned category, as well as a very few metal airplanes that meet the requirements

The type of airplane that the sport enthusiast is most often interested in is the single-engine, two to four passenger, fabric-covered airplane in the 65 to 165 horsepower range that costs between $4000 and $6000 to buy and $800 to $1500 to recover and bring back to good serviceable condition. The budget-minded sport pilot is not interested in newer production aircraft that sell for $15,000 to $100,000, nor is he or she interested in, or able to afford, expensive antique, classic, or biplane machines (Figs. 2-4 through 2-6).

In a nutshell, the choice can be narrowed down to the Aeronca, Piper, Taylorcraft, Stinson, and Cessna type airplanes. There are other possibilities, but these offer the most advantages in terms of

Fig. 2-1. Small local airports are usually the haven of the small plane that may need restoration, but may be purchased at a "bargain basement" price.

easy availability, predictable maintenance requirements, and parts availability, as well as low cost of purchase, restoration, and operation.

I will list here some representative examples of this type of airplane. You can then shop around, compare figures for cost and performance, explore other possibilities not mentioned here, and then make your choice, based on your own needs and desires in a small personal airplane.

Taylorcraft BC 12D

The Taylorcraft BC 12D (Fig. 2-7) was designed and built by Gilbert Taylor, original designer of the Cub. His Alliance, Ohio manufacturing plant turned out about 3000 Taylorcraft airplanes between 1941 and 1947. Today the Taylorcraft F-19 is again being produced in a plant in Alliance, Ohio. The airframe is basically unchanged, though the aircraft is now powered by a Lycoming 0-235 engine of 110 horsepower instead of the 65 horsepower Continental of the earlier BC 12D (Fig. 2-8).

The Taylorcraft BC 12D seats two, side-by-side, and will fly an honest 100 mph on 65 horsepower, thus outpacing by a considerable margin the Cubs and Champs that were its major competitors and contemporaries. Its elliptical airfoil lends itself better to inverted flight than does an airfoil that imparts no curve to the bottom surface of the wing. (Many clipped-wing T-crafts make the airshow circuit every year; the wings on these are a shorter, hot-rod version of the original Taylorcraft.)

As a fun airplane, the T-craft is hard to beat and many consider it easier to recover than others of the type, due to its small, narrow, tapering fuselage with straight lines and smaller surface areas.

Though their performance is slightly better than others, the T-crafts often appear on the used market in various states of repair at a lower price than any of the other popular two-place fabric-covered singles. They can be a real bargain, provided that care is taken to select one that is, indeed, restorable (Table 2-1).

Stinson 108

The Stinson 108-2 "Flying Station Wagon" is a four-place contemporary to the Cub, Aeronca Champ, and Taylorcraft types of the

Fig. 2-2. Old airplanes that have been neglected may look awful; bird nests, beehives, etc., may have been built in every nook and cranny, but hard work and a few dollars may put you in the air in your own plane.

Fig. 2-3. Well maintained sport planes, such as this Aeronca 7AC usually bring between $5500 to $8000 if no restoration is needed, but considerably less if new fabric or an engine overhaul would be required to make the airplane airworthy.

immediate post WWII period. They are sturdy, comfortable, and docile airplanes that have a reputation for honest performance and good handling characteristics (Fig. 2-9).

Something over 5000 of the Stinson 108s were built between 1946 and 1949. The Stinson 108 design was sold to Piper in 1948, but very few were produced by Piper before the airplane was discontinued in 1949.

The Stinson 108-2 was powered by a Franklin 6A4-165 engine of 165 horsepower. Though the Franklin engine company is not currently producing engines, parts for the engines are available still from various sources. One caution, however: there were two types of Franklin engines manufactured, one with a light case and one with

Fig. 2-4. It is probably wise for the first-time rebuilder to avoid rare, antique or specialized airplanes, particularly those with "round" (radial) engines. They can be very expensive to restore because of parts availablity.

Fig. 2-5. When you find an airplane like this classic old Porterfield, it usually means that, with the necessity of engine repair or overhaul, the owner may have lost interest in the airplane. Sometimes these birds represent a bargain to the rebuilder, but care must be taken to accurately assess the cost of putting the ship back in flying condition. An overhauled or factory exchange engine may cost $3500 to $5000.

Fig. 2-6. The Porterfield is less numerous but may also be a good restoration project.

a heavy case. It is economically, if not totally, impossible to overhaul a light-case Franklin 165 horsepower engine. So it is imperative, when contemplating the purchase of a Stinson, to make sure that it has the heavy-case engine.

A distinguishing feature of the 108 is that it is configured with leading-edge slots in the outer portion of the wing that enable the wingtip to maintain airflow much longer before stalling. Stalls in the

Fig. 2-7. The Taylorcraft BC-12 is an inexpensive, fast and fun airplane. This is a particularly well-done restoration.

Fig. 2-8. This is a brand-new Taylorcraft F-19. They are being made again in Alliance, Ohio. The airplane is clearly the property of Gilbert Taylor's original BC-12 design, but the new model is powered by a 110 horsepower engine and costs around $20,000 (courtesy of Taylor Aviation Corp.).

Table 2-1. Taylorcraft BC 12D Specifications.

Engine	Continental
Horsepower	65 hp
TBO	2000 hr
Wing Span	36 ft
Height	6ft 6 in
Empty Weight	750 lbs
Gross Weight	1,200 lbs
Useful Load	450 lbs
Wing Loading	6.5 lbs/sq ft
Power Loading	18.5 lbs/hp
Baggage Capacity	40 lbs
Maximum Speed	100 mph
Cruise Speed	95 mph
Stall Speed	38 mph
Rate of Climb	900 fpm
Service Ceiling	15,000 ft
Range	300 sm

Stinson are, therefore, very docile and the airplane handles very well at low speeds.

Overall, the Stinson Station Wagon is a great four-place family airplane that will return much service for the dollar. It will cost a little more than the smaller craft to restore, and it will consume eight to ten gallons of 80/87 octane fuel per hour (Table 2-2).

Fig. 2-9. The Stinson 108 is a good airplane, though this particular example may be too far gone to be economically restored.

Piper Tri-Pacer

The Piper Tri-Pacer was one of the most popular four-place airplanes of the 1950s. It was actually a tricycle gear model of the popular PA-20 Pacer, with which it shared nearly everything but the nosewheel (Fig. 2-10).

These are plentiful on the used airplane market, though at a price approaching the upper limits of the budget outlined earlier in this book. It costs around $5500 to $9500 for one in flyable condition, perhaps a bit less for an airplane in need of restoration and recovering. However, for the sport flying enthusiast who needs a four-place airplane, the Tri-Pacer is one of the few bargains available for anywhere near the stated budget.

The Tri-Pacer is an easy-to-fly, relatively low cost-of-operation airplane. It will burn from 6.5 to about 9 gallons of fuel per hour, depending upon whether it is equipped with a 125, 150, or 160 horsepower engine. The latter also requires 100/130 octane aviation fuel which is slightly more expensive than the 80/87 octane fuel used by the lower-powered Tri-Pacer and nearly all of the two-place planes discussed herein (Table 2-3).

Piper J-3 Cub

The Piper Cub is an airplane that has become synonymous with light single-engine airplanes to the general public. Long one of the most popular small airplanes, the familiar bright yellow airplane with

Table 2-2. Stinson 108-2 Specifications.

Engine	Franklin 6A4-165
Horsepower	165 hp
TBO	1,600 hr
Wing Span	34 ft
Height	7 ft 6 in
Empty Weight	1,320 lbs
Gross Weight	2,400 lbs
Useful Load	1,080 lbs
Wing Loading	15.5 lbs/sq ft
Power Loading	14.5 lbs/hp
Baggage Capacity	100 lbs
Maximum Speed	133 mph
Cruise Speed	130 mph
Stall Speed	62 mph
Rate of Climb	580 fpm
Service Ceiling	15,000 ft
Range	550 sm

Fig. 2-10. The Piper PA-22-150 Tri-Pacer is a popular and inexpensive four-place, fabric-covered airplane.

the black lightning strip is still going strong, both on the used market and as the Super Cub, which Piper built through 1981.

Starting out as a Taylor E-2 Cub and J-2 Cub, the original 40 horsepower J-3 was produced by Piper after William Piper, Sr., and Gilbert Taylor dissolved the partnership that fostered the design. Taylor went on to design and produce the Taylorcraft B12 and others, while Piper made the Cub a household name.

Powered for most of its years of production by the Continental 65 horsepower engine, the combined total production of the Cub, including the PA-11 Cub Special and the J-4 Cub Coupe, is something over 10,000. The Cub was built from 1938 until 1949, when the PA-18 Super Cub was introduced, the Super Cub being, for the most part, a higher horsepower version of the same basic design.

Since there were so many manufactured, there is no shortage of good used Cubs on the market today. Also, the supply of J-3 Cubs in need of restoration is significant, but both the well-cared-for Cubs and the restoration projects are likely to be more expensive than a comparable vintage Aeronca, Taylorcraft, Porterfield, or Cessna. The prices for a good flyable Cub start at around $8000, and even a reasonable restoration project will cost $5000. The airplane has become, to employ a somewhat hackneyed phrase, a "classic." Thus, the price is high.

Restoring a Cub is no easier nor more difficult than restoring any of its contemporaries, but the bottom line cost will put the finished Cub project in the $8000 to $10,000 ballpark. You will have a beautiful piece of aviation history and a lot of fun when you are finished, though (Table 2-4).

Luscombe Silvaire

The Luscombe series of all-metal taildraggers have become almost a cult airplane in and of themselves. They inspire a fanatic following among those pilots who seek an inexpensive, economical airplane with great handling qualities and the image of bravado that surrounds any Luscombe pilot who successfully handles this sometimes touchy little airplane (Fig. 2-11).

Produced over a long period, from 1937 to 1959, the Luscombe design features a two-place, side-by-side cabin, stick controls, snappy maneuverability, and a rather narrow landing gear that occasionally contributes to a great deal of excitement on the landing rollout. Groundloops are not uncommon in this bird. However, with ample respect for the idiosyncrasies of the airplane, they can be easily mastered.

Most of the Luscombes were powered with Continental engines, ranging from the 65 horsepower version to the C-90.

Buyers should be careful to thoroughly check for stress damage, as these airplanes, though not specifically certified as aerobatic, have long been flown through aerobatic manuevers. They have such a nice snap-roll and aileron roll talent that it is difficult for pilots to resist the temptation. They are also easily looped, particularly the higher horsepower versions, and if not handled proficiently, can fall out of aerobatic maneuvers as easily. So it is wise, when contemplating any Luscombe purchase, that your mechanic make a thorough inspection to detect any structural problems.

Table 2-3. Piper PA-22-150 Tri-Pacer Specifications.

Engine	Lycoming O-320
Horsepower	150 hp
TBO	2,000 hr
Wing Span	29 ft 4 in
Height	8 ft 4 in
Empty Weight	1,100 lbs
Gross Weight	2,000 lbs
Useful Load	900 lbs
Wing Loading	13.5 lbs/sq ft
Power Loading	13.3 lbs/hp
Baggage Capacity	100 lbs
Maximum Speed	137 mph
Cruise Speed	125 mph
Stall Speed	48 mph
Rate of Climb	750 fpm
Service Ceiling	15,500 ft
Range	500 sm

Table 2-4. Piper J-3 Cub Specifications.

Engine	Continental
Horsepower	65 hp
TBO	2,000 hr
Wing Span	35 ft 2 in
Height	6 ft 8 in
Empty Weight	680 lbs
Gross Weight	1,160 lbs
Useful Load	480 lbs
Wing Loading	6.18 lbs/sq ft
Power Loading	22 lbs/hp
Maximum Speed	85 mph
Cruise Speed	70 mph
Stall Speed	38 mph
Rate of Climb	450 fpm
Service Ceiling	12,000 ft
Range	200 sm

Like other metal airplanes discussed herein, some of the Luscombes were painted, though many owners desire to obtain the brilliant all-metal luster that distinguishes most examples of this craft. Beware of the painted model, since it is much more difficult to achieve the beautiful metal luster once the airframe has been painted (Table 2-5).

Aeronca Champion

The Aeronca Champion model 7AC is a two-place tandem (front and rear) seat high-wing monoplane powered by a four-cylinder Continental 65 horsepower engine. The airplane was first produced in 1945 and various versions have been produced since (Figs. 2-12, 2-13).

In 1947, the airplane appeared with a Continental 65 horsepower engine, and until recently the Bellanca Aircraft Corporation built the same basic Champion airplane with 115 and 150 horsepower engines.

There were around 7200 of the Champion airplanes built, and a good number of them are still flying and are available on the used market. Prices vary from $2500 for a non-flyable restoration project to $9000 with new fabric and zero time since major engine overhaul. These are excellent basic flying machines; they are among the lowest-cost airplanes to buy and their miserly little 65 Continentals burn a mere 4.2 gallons per hour and have a TBO (time between overhauls) or expected service life of 2,000 hours (Table 2-6).

Ercoupe/Aircoupe

The Ercoupe is a low-wing monoplane two-seater, powered variously by engines ranging from the Continental 65 horsepower engine to the C-90. It is unique in that, for most of its 30-odd year production life, the aircraft was billed as non-spinnable and built without rudder controls, the ailerons being tied in with the rudder cables. The aircraft was controlled around all axes by the control wheel. This made the airplane ultra-simple to fly, but at times difficult in crosswind situations, since cross-controlled flight, slips were impossible; it was considered by many to be a joke of the industry. Nonetheless, around 6300 of the little ships were built and sold by such companies as ERCO (Engineering Research Corporation), Forney Aircraft Company, Alon Aircraft Company, and finally, the Mooney Aircraft Company (Fig. 2-14).

A large number of these airplanes are still flying. They are simple and sporty, though not impressively fast. My own opinion is that the craft have a breathtaking glide ratio, akin to a grand piano, but are very nice due to their canopy-type entry/exit system that can be opened in flight for the open-cockpit effect. (Make sure your seat belt is securely fastened!)

Fig. 2-11. The all-metal two-placers such as the sporty and challenging little Luscombe Silvaire demand much elbow grease to maintain the high luster. This one is so highly polished that it can easily be used as a mirror for shaving on long cross-country trips.

Table 2-5. Luscombe Silvaire Specifications.

Engine	Continental
Horsepower	65-90 hq
TBO	1,600-2,000 hr
Wingspan	35 ft
Height	6 ft 3 in
Empty Weight	800 lb
Gross Weight	1,200 lb
Useful Load	500 lb
Wing Loading	9.2 lb/sq ft
Power Loading	14.4 lb/hp
Baggage Capacity	80 lb
Maximum Speed	130 mph
Cruising Speed	120 mph
Stall Speed	40 mph
Climb Speed	70 mph
Rate of Climb	900 fpm (90 hp)
Service Ceiling	17,000 ft
Range	500 sm

Though some later models were equipped with rudder pedals, most were not, and it is a little hard for a pilot with "conventional" airplane experience to get used to the idea of "driving" the airplane around in the sky with feet flat on the floor. Another distinctive feature is the wide bench seat, which makes the older Ercoupes look and ride sort of like a Model A club coupe. Overall, these airplanes are inexpensive—they can be purchased, even today, for around $4500, even in good condition (and, of course, less when in need of restoration). They will require a minimum of maintenance and will consume in the neighborhood of 5 to 6 gallons of fuel per hour. Earlier models had fabric-covered wings, which will eventually need cover, but this process, owing to the small amount of fabric involved (compared to fully fabric covered airplanes) is straightforward and rather inexpensive. As with all metal airplanes, be careful of wrinkles in the metal skin and other evidence of metal damage and/or fatigue (Tables 2-7 and 2-8).

Piper Colt

Piper produced this slightly smaller, lower horsepower airplane as a two-place version of the PA-22 Tri-Pacer. Somewhere around 2000 of the model were built from 1960 to 1962 when Piper discontinued the fabric covered PA-22 designs in favor of the sleeker, more modern looking PA-28-140 Cherokee.

Fig. 2-12. The Aeronca 7AC Champion is one of the most popular of the older two-place lightplanes.

Fig. 2-13. Perfectly flyable light airplanes such as the Aeronca 7AC Champion show up for sale at prices in the $4000—$6000 range.

Table 2-6. Aeronca 7AC Champion Specifications.

Engine	Continental
Horsepower	65 hp
TBO	2,000 hr
Wing Span	35 ft 2 in
Height	7 ft
Empty Weight	809 lbs
Gross Weight	1,300 lbs
Useful Load	491 lbs
Wing Loading	7.63 lbs/sq ft
Power Loading	15.3 lbs/hp
Baggage Capacity	40 lbs
Maximum Speed	102 mph
Cruise Speed	90 mph
Stall Speed	32 mph
Rate of Climb	65 mph
Service Ceiling	14,700 ft
Range	250 sm

The Colt, powered by the Lycoming 0-235 engine of 108 horsepower, is a pleasant and straightforward fun airplane (Fig. 2-15). It is easy to fly and reasonably comfortable with two-abreast seating. It is an all-fabric airplane, and is easy to recover and/or restore to its original sporty look.

Fig. 2-14. The Ercoupe was produced for many years under several names, Forney, Alon, Erco and Mooney. Some had rudder pedals, but most were simply controlled by the wheel, which interconnected ailerons with rudder. The unique-looking airplane is loads of fun and cheap to maintain and fly.

Table 2-7. Ercoupe 415-G Specifications.

Engine	Continental
Horsepower	65-85 hp
TBO	1,600-2,000
Wing Span	30 ft
Height	5 ft 11 in
Empty Weight	838 lb
Gross Weight	1,400 lb
Useful Load	562 lb
Wing Loading	9.8 lb/sq ft
Power Loading	16.5 lb/sq in
Baggage Capacity	75 lb
Maximum Speed	125 mph
Cruising Speed	110 mph
Stalling Speed	48 mph
Climb Speed	75 mph
Rate of Climb	560 ft per min
Service Ceiling	11,000 ft
Range 1	450 sm

As was the case with nearly all the two-place airplanes we are dealing with, the Colt was primarily a trainer, and as such, was subjected to occasional hard use. A thorough check of the aircraft logbooks should be made to determine whether or not the aircraft has ever been seriously damaged, and, if possible, how long it spent in service as a trainer. Fortunately, with the introduction of the

Table 2-8. Alon Aircoupe A2 Specifications.

Engine	Continental
Horsepower	90 hp
TBO	2,000 hr
Wing Span	30 ft
Height	5 ft 11 in
Empty Weight	838 lb
Gross Weight	1,400 lb
Useful Load	562 lb
Wing Loading	9.8 lb/sq ft
Power Loading	16.5 lb/hp
Baggage Capacity	75 lb
Maximum Speed	125 mph
Cruising Speed	110 mph
Stall Speed	48 mph
Climb Speed	75 mph
Rate of Climb	560 ft per min
Service Ceiling	11,000 ft
Range	450 sm

Fig. 2-15. The Piper PA-22-108 Colt, a two-place version of the Tri-Pacer.

Cherokee trainers simultaneous with the manufacture of the last of the Colts, some Colts were turned out of flight school service rather early and may not have seen much, if any, duty as trainers. Service as a flight school machine should by no means cause you to rule out any airplane, but it *is* cause for further scrutiny into the history of the ship (Table 2-9).

Cessna 120/140

The Cessna 120 and 140 series are two-place airplanes with side-by-side seating. The Cessna 120 features a metal fuselage, and when new, had fabric-covered wings. The 140 was all metal and was equipped with an additional rear window and wing flaps. Many of the 120s were modified with the window conversion and metalized wings, making them more like the 140, particularly when the standard Continental 65 horsepower engine was replaced with the Continental C-85 or C-90.

The little Cessnas were well appointed inside, with a nice-looking instrument panel and control yokes, but are a bit narrow, making flight by two persons rather cozy, to say the least.

Restoring an example of this mark probably will involve many hours of polishing the metal fuselage, as the majority of 120s and 140s bore only trim paint on their fuselages.

Beware of painted metal surfaces if your goal is to have a bright, polished metal airplane. Paint has a tendency to etch into the metal, making subsequent attempts to return the original metal luster very difficult.

Also, make sure that your mechanic makes a careful inspection of the landing gear attachment points, and the fuselage bulkheads, as these places are vulnerable to damage during hard landings, groundloops, and other accidents; since the Cessna 120 and 140 demand quite a bit of attention of a pilot on landing, the airplanes are more susceptible to such occurrences.

Performance-wise, a lot of positive things are attributed to the 120/140 design. Even with 65 horsepower, the airplane is respectively fast—around 90 mph. With the C-85 or 90, the ship is good for over 100 mph and will keep up with any Cessna 150. The older 65 horsepower models do take quite a distance on takeoff and landing, however. Fuel consumption is a frugal 4-6 gallons per hour. Available in the $5000 range when in need of restoration (possibly less), these airplanes represent a good investment, as appreciation is the rule with them (Table 2-10).

All-Metal Airplanes

All-metal airplanes merit some special considerations. Care must be taken to insure that these airplanes are free of rust, corro-

Table 2-9. Piper PA-22-108 Colt Specifications.

Engine	Lycoming 0-235
Horsepower	108 hp
TBO	2,000
Wing Span	30 ft
Height	6 ft 3 in
Empty Weight	940 lbs
Gross Weight	1,650 lbs
Useful Load	710 lbs
Wing Loading	11.2 lbs/sq ft
Power Loading	15.3 lbs/hp
Baggage Capacity	100 lbs
Maximum Speed	120 mph
Cruise Speed	115 mph
Stall Speed	55 mph
Rate of Climb	600 fpm
Service Ceiling	12,000 ft
Range	234 sm

Table 2-10. Cessna 120/140 Specifications.

Engine	Continental
Horsepower	65-90 (120-140)
TBO	1,600-2,000
Wing Span	33 ft 4 in
Height	6 ft 5 in
Empty Weight	890-900 lb
Gross Weight	1,450 lb
Useful Load	485-560 lb
Wing Loading	9.1 lb/sq ft
Power Loading	17.1 lb/sq ft
Baggage Capacity	50 lb
Maximum Speed	125 mph
Cruising Speed	105 mph
Stall Speed	45 mph
Rate of Climb	600 ft/min
Climb Speed	80 mph
Service Ceiling	15,500 ft
Range	450 sm

sion, and internal structural damage. This is where a good, qualified mechanic is a must. But once a clean aircraft is selected from this group, the fun of flying can go on, even as you scrub, polish, and tinker with these airplanes without tearing them completely down for an airframe overhaul.

On the negative side of the coin, however, is the fact that these airplanes may not be quite as cheap to fly and maintain, and, of course, the considerable savings and satisfaction of buying at a very low price and restoring to mint condition may not be realized to the extent that would occur with a fabric bird.

This should give you a good idea of what is generally available in affordable small airplanes. Once the choice of a type is made, then the search can begin for one that meets the requirements of general condition and price that will fit your needs the best. You are going into the project knowing that you are going to trade off labor for a low purchase price, so the physical condition of the airframe is of less overall importance than the purchase price, considering the amount you will have to spend on restoration. There is one other very large consideration that will be discussed in the following chapter, one that you as a non-mechanic can do little work on, and that's the engine. Much care will be required to make a good choice on an airplane with an engine that will last long after the restoration project is complete.

Chapter 3
Lightplane Engines

Naturally, one of the biggest considerations in buying and owning an airplane is the engine, its reliability quotient, and its expected remaining service life. It is important to know some of the basic facts about small aircraft engines, their performance, required service, and reliability or expected service life before overhaul.

In all probability, you will be selecting a light aircraft with between a 65 and a 165 horsepower engine. It will either be a four or six cylinder engine and will probably be manufactured by either Lycoming or Continental (Fig. 3-1). In a few cases, you may come into contact with a Franklin powerplant in some of the Taylorcraft and Stinson models.

The Continental 65 horsepower engine has been very popular on light aircraft ever since the 1940s. Therefore, the vast majority of light two-place aircraft are powered by this very fuel-conserving and reliable engine. (Fig. 3-1). The 65 Continental powers the Piper J-3, the early Taylorcraft BC 12 series, the Aeronca 7AC, the Luscombe Silvaire and others. It is a four cylinder horizontal-opposed engine that burns 80/87 octane aviation fuel and can, with proper care and maintenance, be expected to operate for 2000 hours in the air. Both the Lycoming 0-320 and Lycoming 0-235, which are commonly found on the Piper Tri-Pacers and Piper Colts respectively, are also four cylinder engines, and can also be expected to operate 2000 hours before a major overhaul. The six-cylinder Franklin 6A4-165 engine found on the Stinson 108-2 has a somewhat lower expected service life between overhauls of 1200 to 1600 hours. (The time-between-overhaul figure is established by the manufacturer of later model

31

Fig. 3-1. The four-cylinder Continental 65 has long been one of the most popular and reliable light airplane engines.

powerplants as a recommended time at which the engine would be dismantled and overhauled. The early engines were built prior to the time when manufacturers recommended a time between overhaul.)

It is important to remember that the amount of time between overhauls is directly dependent on the quality of maintenance that the engine has seen throughout its service life and the care with which it was flown. Sometimes during an inspection, the engine compression test reveals a weakness of one kind or another in the engine, it will be taken partially apart for a top overhaul. This means that the rings are replaced and, possibly, that the cylinders are honed and all parts checked for conformance with established wear tolerances and that the engine is reassembled without taking the crankcase apart. This superficial overhaul might then enable an engine that had begun to use an inordinate amount of oil, that failed its annual compression check or was in some other way not running properly, to reach a much higher time before the complete overhaul is necessary.

Engine Logs

When looking at a potential aircraft purchase, carefully study the engine log to determine how the engine has been maintained, what the compression test figures were at each inspection conducted since the last major overhaul, and, in general, what work had to be done to it. This history of service of an engine will tell you much about what to expect of the powerplant in terms of service and reliability. An engine that has exhibited regular and/or chronic problems will no doubt continue to plague its owner with further costly trouble, not to mention the safety consideration of a questionable engine. On the other hand, an engine that has turned out several hundred hours of satisfactory and reliable service, that has been regularly and properly maintained and serviced with regular oil changes, etc., can usually be depended upon to continue to give reliable service.

When studying the aircraft engine logbook, there are certain critical things to analyze. These are the differential compression numbers, recorded at each annual and/or 100 hour inspection and any data noted in the engine log pertaining to the inspection of oil screens that might indicate the presence of metal particles in the oil. Normally, the differential compression check involves removing the spark plugs from each cylinder and inserting a plug attached to a compressed air source. Most aircraft cylinders are tested with 80 pounds per square inch of air pressure. Each cylinder should hold a good percentage of this 80 psi value. Over 70 psi is a good rule of thumb. The difference between the air pressure applied to the cylinder and the amount indicated on the compression tester gauge is the amount of air that is finding its way past the piston rings. Since compression is a direct indication of the power output potential of the internal combustion engine, compression is a direct indication of the amount of wear that has occured inside the engine. A newly overhauled engine should test at 75/80 on all cylinders. As the engine time builds and wear continues, the compression test figures or the ratio between the two will gradually diminish. Once the figures in the logbook begin to indicate test figures in the 50 to 60 over 80 range, you can begin to expect that major maintenance will soon be required.

Occasionally, there will be information in the log indicating that an inspection revealed one cylinder to be below the required compression. This situation calls for repair on one or more cylinders. When a repair is made to bring the one cylinder up to the required compression, a logbook entry may be made, indicating that the single cylinder received a top overhaul. Provided that the cylinder then checks within the required compression value, the engine is adequately repaired.

With respect to the inspection of oil filters and/or screens, beware of any engine that is said to have a lot of metal in the oil screens. This indicates a chronic condition inside the engine, and the need for immediate work. A *small* amount of small particle metal detected in the screen is considered normal, inasmuch as the engine is undergoing a small, expected amount of wear at all times. However, any particles that can visibly be picked up with a magnet should cause concern and an immediate investigation into their source.

Top Overhaul

A top overhaul will cost approximately $150 to $200 per cylinder, as a general rule of thumb, but once properly done, a top overhaul could enable you to obtain the maximum service life from the engine.

You may discover that an engine has had several top overhauls, every 300 to 500 hours. This does not necessarily mean that it is a *bad* engine, but it might clue you in to the way in which the engine was operated, which of course, indirectly results in the length of service life of the engine. An engine that requires numerous top overhauls has probably done a lot of slow climbing on hot days, possibly in flight school service, followed by rapid cooling in descents. Such treatment shortens engine life and results in a much shorter interval between required major engine maintenance.

Major Overhaul

You may also be faced with an engine that has zero time since major overhaul. Theoretically, if the overhaul was done by a qualified mechanic using procedures approved by the manufacturer and the FAA, this newly major overhauled engine should be almost as good as new. During a major overhaul, the entire engine is taken completely apart, and each part carefully checked to ensure that it meets service tolerances. Any part not meeting these close tolerances is replaced with a new part. Properly done, this results in a virtually like-new engine. Such an engine can then be expected to operate for the same length of time that a new engine would, perhaps 2000 hours plus.

Basically, the less time since major overhaul, the more the engine is worth, and the more it will add to the price of the airplane. If you are buying an airplane that appears to be cheaply priced but has a high time engine of say 1800 hours, or one that has been sitting outdoors neglected for years, you are going to be faced very soon with the need for a major overhaul. Bear in mind at all times during your search for an inexpensive aircraft that the current cost at this writing for a factory remanufactured or freshly overhauled engine is between $3500 and $5000 depending on the size of the engine. For this purpose, I am speaking of overhauled Continental and Lycoming engines of 65 to 165 horsepower, the only kind that are readily available. Also beware of shops or individuals that offer engines with zero or very low times since overhaul, for they may have been done in questionable shops. If you are looking for a newly overhauled engine, or a shop that can overhaul your engine, be very careful to locate a shop that is both FAA licensed and reputable to do the type of work you need done.

All things considered, it is best to buy an airplane that has an engine with no more than 1000 hours on it since a major overhaul. This would be referred to as a mid-time engine for most light aircraft, and since you will probably fly no more than 300 to 500 hours per year, this engine could be expected to give you a minimum of two or three years of good service. Airplanes with engines in the zero to 700 hour range will cost you more, but will be good for much longer before major engine maintenance is required, and therefore will be more valuable at resale time.

When discussing the cost of engine maintenance and overhaul and the effects of these factors on initial cost and resale value, it is an excellent time to remind the buyer that the *real* cost of owning an airplane is not only what you pay for the airplane and how much it costs you to fly and maintain it, but how much you are able to sell it for when it is time to trade. If you buy at a reasonable price and restore the airplane, you can realize a nice appreciation on your investment. For example, an aircraft bought for $4000 that has a mid-time powerplant will, when recovered with $700 to $1000 worth of materials, be worth between $7000 and $10,000, an adjusted appreciation of $2300 to $5000!

Looking at the purchase from this standpoint, it is conceivable that you can refurbish a light airplane, fly it for a couple of years, and still make a nice return on your initial investment. This situation is bound to improve even further as the costs of fuel and new production airplanes increases, driving the pleasure of private flying further and further beyond the reach of the average person.

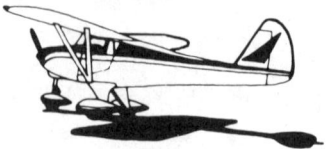

Chapter 4
The Pre-Purchase Checklist

Before buying any used airplane, you should give it a flight test. This is the only way you can determine the handling characteristics, trim, and actual engine operation. This is another good reason for not buying a completely run-out airplane that is not airworthy at the time of purchase. In some cases, you might be attracted to a low-priced airplane that has simply gone out of license for lack of an annual inspection and that is, therefore, not flyable and low in price. Often a good purchase can be made this way, but by the same token, there is no substitute for the added peace of mind afforded by at least having flown the airplane through a brief check to make sure that it *does*, indeed, fly.

If an airplane is airworthy when you are contemplating buying it and has a current airworthiness certificate and annual inspection, you should be allowed to fly the airplane. Be wary of any airplane that the owner will not let you fly. If the owner does not feel comfortable about allowing you to fly the airplane alone, get him to go along, even to make the takeoffs and landings if necessary, or if necessary, hire an instructor that is agreeable to both you and the owner, to go along. If the owner refuses to allow you to fly his airplane for no apparent reason, decline the purchase. It may mean that the airplane doesn't fly well, and that usually means that the airplane is either badly out of rig (usually a correctable problem that you would solve during the rebuilding process anyway) or it may be sprung from a crash. Airplanes that have been sprung, or have had their fuselages warped somehow, don't fly very well, and there sometimes is not a whole lot that can be done about it. Even if the fuselage can be straightened, it is a complicated process, involving putting the fuselage in a jig and precisely realigning it. Stay away from this if at all possible. It's not worth it.

Before flying any unfamiliar airplane, thoroughly go over what you'd like to do with the owner or second pilot. Assure him that you are not interested in "wringing it out," only conducting a routine test. Then do a very thorough preflight, as this can also tell you a great deal about an airplane.

Cockpit

Start with the cockpit. Make sure that the switches are off. Remove any control locks that are installed, and move the controls through their full travel. They should move freely and easily, with no catching or rubbing. Stiff controls can mean poor lubrication, frayed control cables, bad pulleys or obstructions to the control travel inside the airframe. Look out at the control surfaces and ascertain that they are moving freely and in the correct direction.

Wings

After the cockpit check, begin at the wing robot and check the condition of the wing attachments, if they are accessble through inspection plates or fairings. Don't be afraid to remove inspection plates—remember that you are considering putting out a few thousand hard-earned dollars here, and you need to see all you can. If when you open an inspection cover you are horrified by what's inside, i.e., bird's nests, rust, or corrosion, you'll want to at the very least take this into consideration when deciding whether or not to buy; moreover, you may want to reconsider at that time whether or not you wish to continue the preflight and fly the airplane!

Continue down the wing, looking into the inspection covers, into control channels, and at the attachment points of struts and wires. Also check the main landing gear for inflation and hydraulic leaks, if the airplane uses hydraulic brakes. Check to make sure that the drain grommets on fabric-covered airplanes have not closed up with dirt, grime, and/or mildew. This could mean that moisture would be trapped at the trailing edge, instead of draining out, and that could mean trouble for the wood or metal inside. If the drain grommet opening is not clear, you may see a brown-orange substance filling it. This is dust and mildew. Not necessarily a disqualifying factor, since all older airplanes are likely to accumulate some over the years, but if *all* grommet holes are plugged, further investigation is recommended. Check the aileron attachment fittings and the freedom of movement of the ailerons. Look for rust and corrosioin at the attachments and control fittings. Finally, check over the wingtip and leading edge. The tip should be free of damage, and the leading edge should be relatively smooth. A badly dented leading edge will cost you an extra $250-350 in the rebulding process.

Engine

As you move around the nose of the airplane, open up the cowling and take a good look at the engine. Check for loose wires and hoses. Be especially aware of oil leaks or evidence of oil leaks around the bottom of the crankcase and at the base of the cylinders where they join the crankcase. I stress evidence of oil, because, in all probability, the seller has thoroughly washed down the engine and the inside of the engine compartment prior to showing you the airplane, particularly if there was any major oil leak. If oil has been leaking onto an engine, it will usually stain it so that it will be visible to the careful observer. If you have your mechanic with you, as is strongly recommended, he will be able to help you determine the extent of oil leaks, and it is also strongly recommended that he performs a differential compression check on the engine at this point. This will tell you a lot about the condition of the engine and its life expectancy.

Check the oil, and while doing so, see how black it is. Aircraft engines, since they operate mostly at altitude and not down in the exhaust fumes like automobiles do, should operate much cleaner, and thus the oil should remain fairly clean from one oil change interval to the next. It might be dark brown toward the end of an interval but should not be pitch black like automobile oil is when you change it. Ask the owner how often the oil is changed. It should be done at no more than 25-hour intervals. Again, if he's trying to sell the airplane, he has probably just changed the oil.

Upon closing up the engine compartment, move around to the propeller. Check it over carefully for nicks, cracks, and bends. A propeller can tell you a lot about the care that the airplane has had and about the airports from which the airplane has been operated. A ragged, gouged, or bent propeller is trouble and might tell you that the owner was not careful about his airplane or its maintenance. Any nicks will have to be filed out by an A&P mechanic, and a bent propeller will probably have to be overhauled by an FAA approved prop shop. *Never try to straighten a bent prop yourself!* Pull on the prop with your hands close to the hub. It should not be overly loose on the shaft. Then, after making sure that the electrical switches or the magneto switch is off, carefully turn the propeller through a few times and pay close attention to what you feel. Is it stiff? Does it simply turn over freely with very little effort? An engine with proper compression should require some effort to turn over and should snap through the compression stroke. Again, a compression check is the real test of the condition of the engine, but a very weak engine will let you know when you pull it through.

While at the front of the airplane, check the cowling for cracks and tears and also check the oil cooler intake and air intakes for obstructions.

Moving along the other side of the airplane, you will need to check the pitot-static system, if you didn't find it on the other side, and/or the venturi, if equipped. Check these for general condition and for obstructions to their openings. Make all the checks to this wing that you made to the other, then move back to the aft fuselage.

Fuselage

On fabric-covered airplanes, check the fuselage fabric, and all fabric for that matter, for "ring-worm" or cracks in the fabric finish. Your mechanic should make a fabric test, probably with a punch tester. Be sure to get the permission to do this from the owner before proceeding, however, because you will be punching holes in his airplane, and those holes will have to be patched. If the fabric tests out "in the green," that is, in one of the green bands on the punch tester, then the airplane is airworthy and not in immediate need of recovering. If it tests in the yellow band, then the airplane is marginally airworthy and will soon need cover. If it tests in the red, then the aircraft is not airworthy and illegal for flight, and will need new fabric before further flight. Make several punch tests; one in each wing, one on the upper rear fuselage and one on each control surface. The results of the fabric tests will have a direct bearing on the value of the airplane and the amount of work you are going to be faced with upon buying the airplane. Never buy a fabric-covered airplane until its value is determined in part by a test of the essential fabric covering.

While inspecting the fuselage, get down underneath and pull off some or all of the fuselage belly inspection covers. Look up into the rear fuselage with a flashlight, and look carefully at the airframe members, wood formers, control cables and pulleys, and bulkheads. Check for rust on the metal tubing, rotted wood formers, frayed cables, bad pulleys, and corrosion on bulkheads and metal parts. Don't be too surprised if you find some rust, since an airplane that has been covered for years is likely to have a little, but if the metal tubing is substantially unprotected by primer and covered with rust, you may have a problem when rebuilding. This is an inspection that is best done by a qualified mechanic who is capable of judging the extent of rust and corrosion, since this will dramatically affect the value of the airplane and the work it will take to rebuild it.

Tail

Once the fuselage inspection is complete, carefully inspect the tail group. Make sure that the elevator and rudder controls are properly attached and move freely in the proper directions. Check flying cables, struts and wires for rust, corrosion and tautness. Stand back and look at the fuselage and tail group. Does it appear true, or can you see warps and bends that may have been caused by an accident? Slop in the brace wires or cables might also be an indication of an out-of-true airplane. Again, your mechanic is best qualified to make a judgment about questionable areas.

After inspecting the tailwheel on airplanes so equipped, and the other side of the fuselage, you are ready to either conclude your negotiations with the owner, or if still interested in the airplane, take it out for a thorough test flight evaluation.

Flight Test

Assuming that you are pressing on with the flight evaluation, climb aboard the airplane and spend a few moments familiarizing yourself with the layout of the cockpit and the location of all controls, instruments and switches. Move the controls so that you feel the full travel and know where it is. Check to make sure that the airplane has a current airworthiness certificate visible somewhere in the airplane. You should have already gone over all of the logbooks so that you are sure that the airplane has been inspected in accordance with applicable FARs. Then start the engine, after completing the pre-starting checklist.

The engine should start smoothly, and the oil pressure should indicate in the green after no more than 30 seconds. If it does not, shut it down and investigate the cause of the lack of immediate oil pressure.

Immediately after starting the engine and letting it stabilize in a smooth idle, allow it to run for a minute while you listen and feel the airplane. Check for smoothness and vibration.

As you taxi, see how the airplane handles on the ground. Check the brakes immediately after the airplane starts to roll—do not wait until you are halfway down the ramp before finding out if they don't work. Move the controls, checking for freedom and proper direction of movement. Use the pre-takeoff checklist and perform a full run-up.

When you run the engine up, make sure that the carburetor heat works by noting a drop in the rpm upon application of heat. Also make certain that the magneto drop is not excessive. It is also imperative that the rpm drops upon switching magnetos. No drop may indicate a short in the switch, a grounded mag, or a variety of other ills.

Once the pre-takeoff checklist is complete, roll out onto the runway and smoothly apply power. Pay attention to the engine and note any flat spots, roughness or other indications of malfunction. Monitor the engine gauges for any indication of low or fluctuating oil pressure, overheating of oil or cylinder heads, or any major fluctuation in rpm. Climb out at the recommended best rate of climb speed. How does the airplane respond to trim? How does it feel? An airplane should tend to fly straight if properly trimmed and rigged. Once at a safe altitude, say fifteen hundred feet or higher, trim the airplane out for level flight at cruise power.

Once established in cruise and the elevator pressures trimmed off, the airplane should fly straight and level hands-off. If it doesn't but instead wants to peel off to the left or right, you may either have a rigging problem, or something more serious. A mechanic should definitely be consulted on this. The reason I stress this possibility of something other than a mere rigging problem is that I have a friend who once bought an airplane that flew sideways all the time. Thinking it was a rigging problem, he tried everything under the sun to rectify a rigging problem (he is an A&P mechanic). Washing the wings in or out didn't help; indeed, *nothing* helped. The airplane continued to fly sideways. Then, when the time came to recover the airplane, the real problem was discovered: the airplane had been damaged at some time and when it was reassembled, one of the main fuselage beams, where the wing spar attaches to the fuselage was *two inches shorter* than the one on the other side! No wonder the airplane flew like a crab walks! Needless to say, an extensive, expensive repair job, involving much cutting and welding was involved in correcting the problem. This is a strong case for knowing the maintenance history of the airplane *and* having a mechanic check out the bird before purchase.

As long as the airplane flies straight and the engine performs nicely, you can go ahead and do a few turns, perhaps some steep-banked turns and possibly a few stalls, just to get the feel of the airplane. Once you have accomplished all the routine checks you need to satisfy yourself about the airplane, re-enter the pattern for landing—don't burn up excess fuel or get too playful with the airplane (yet); after all, it's not your airplane.

After landing, provided that you and your mechanic both agree that the airplane flies and performs up to standards, and you both agree that the aircraft is worth the asking price, make the deal. After the money changes hands, you will want to make sure that you get the following things: The airplane, of course, along with its Airworthiness Certificate, registration and logbooks. The owner will sign the certificate of registration form, transferring ownership. You will have already run a title search on the airplane, as mentioned in an earlier chapter, to ascertain that the airplane is free of any liens or other encumbrances, and that the owner is, indeed the sole rightful owner of the airplane. A bill of sale form will also be filled out, the pink copy of which constitutes your proof of ownership until the completed registration form clears the FAA and is returned to you. With the paperwork completed, you are ready to fly, or to get started with your restoration project, knowing that you and your mechanic performed a thorough pre-purchase inspection on the airplane, and that you paid no more than the airplane was really worth.

Chapter 5
Long-Distance Buying

Unless you are fortunate enough to have an abundance of small aircraft at an airport near you and are able to make a selection from these, you are faced with a long-distance search that may present some difficulties. This chapter will help you avoid some of the pitfalls of the long-distance search and purchase of a used airplane.

You will want to check all of the airports in your area for airplanes of the type you are interested in that bear "For Sale" signs. Also, check the bulletin boards of the various fixed base operators on the airport. Next, check the classified section of your newspaper, then the classifieds of the newspapers in nearby cities. Should you fail to turn up anything in these searches, you are faced with the need to start shopping the nationwide used-airplane market.

There are a number of periodicals that specialize in used airplane advertising. Among them are the: *A/C Flyer,* published by the Ziff-Davis Publishing Company, Inc. (One Park Avenue, New York, NY 10016; telephone (212) 752-3500); *Aviator's Hotline* (P.O. Box 958, Fort Dodge, IA 50501; telephone (800) 247-2244); and possibly the largest single source of advertisements for the type of equipment we're talking about is *Trade-A-Plane* (Crossville, TN 38555; telephone (615) 484-5137). *Trade-A-Plane* is issued three times per month and is solid ads, cover to cover. The entire newspaper-style magazine consists of classified and display advertising. Virtually everyone who has been in aviation long enough to own an airplane knows about *Trade-A-Plane* as a national medium for exchange for anything having to do with airplanes. It is also the one most likely to contain more than a few ads for the small low-priced airplanes for which you are looking. However, a few words of caution are in order about buying anything on the basis of an ad, anywhere, anytime.

Perhaps the best way to illustrate some of the potential pitfalls of buying long-distance is to relate an experience I had as a first-time airplane buyer.

My wife, who is also a pilot, and I decided that we could no longer live without a small airplane of our own. We had narrowed the choices in an airplane down in the same manner that I outlined earlier, i.e., a two-place, fabric-covered monoplane powered by a 65 to 85 horsepower engine. After discussing Taylorcrafts, Piper J-3 Cubs, Aeronca Chiefs and Piper Colts, we finally decided on an Aeronca 7AC Champ.

The major reasons for choosing the Champ were: (1) there are lots of them available; (2) they are among the lowest priced airplanes on the market; (3) they are simple to restore and maintain, this being a major consideration, since the difference in price between a flyable restoration project and a newly recovered and restored airplane can be as high as $7000; (4) I had learned to fly in a Champ and had quite a bit of experience with them—enough to know a little bit about what to look for and avoid, what added and detracted from the value of the airplane; and (5) the Champ, in addition to its straightforward, low maintenance cost, burns a mere 4.2 gallons of fuel per hour and is thus probably the most economical of them all. So it was that on a cold February Sunday night, we decided that before the next week was out, we would be in possession of our dream plane.

The problem was, however, that there were no Aeronca 7AC Champs available right at that time in our local area, or so we thought. So a long-distance search was started. As a regular subscriber to *Trade-A-Plane*, I grabbed the latest issue and turned to the classified section marked "Aeronca—For Sale." There were fifty or more ads there, about thirty of them were for 7AC Champs of the 1946 to 1947 vintage (the best years in my opinion), and the remainder were for Chiefs, various military versions of the Champion, and a few of the rarer four-place Aeronca Sedans. How could we go wrong from such a selection? What I *should* have asked myself is: just how far wrong *can* we go with this selection? I started calling the telephone numbers listed in the ads, but not before it occurred to me that there was not a single airplane listed that was closer than about a thousand miles from our home. I might point out here that *Trade-A-Plane* receives ads from all over the country, but the majority of these ads are naturally from highly populated areas east of the Mississippi, or at least the Continental Divide. There are fewer from the western U.S.

After making a couple dozen calls and asking many probing questions, I found out a lot more things about the airplanes advertised than was to be found from the ad itself. For example, an owner advertising an airplane for sale might not indicate in the ad that the airplane has been slightly wind damaged and needs some repair before it can be flown. Another might neglect to mention that the airplane has an engine that is very high time and in need of a major overhaul. Most ads list the total airframe and time since overhaul. Some ads do not. By the same token, some ads fail to mention price. In calling on these ads, I frequently discovered that an owner seeking to sell a more-or-less average airplane was asking a price that was several hundred dollars or more above the average market value for the airplane. And so it went, until I finally found an airplane that sounded to me to be ideal, not only from the sound of the ad but also from talking to the owner as well.

The airplane was an Aeronca 7AC Champion with a Continental 85 horsepower engine. It had zero time since major overhaul on the engine and what sounded like a fairly recent fabric job. The color sounded very nice, the airplane was owned by a doctor (which always inspires confidence), and it was available for a price well within our price range of $4000 to $6000. This was on Wednesday, and after a brief discussion, we decided that I would travel out to the midwest to pick up the airplane on Friday, if the title search went okay and the airplane was all it was advertised and claimed to be by the owner.

The following day was spent in running a title search through the Aircraft Owners and Pilot's Association (AOPA) title search office, located next to the FAA headquarters in Oklahoma City. The title was clear, meaning that no liens or encumbrances existed against the airplane. I called my insurance company, had a binder put on the airplane for the flight home, and made airplane reservations (one way) to the city nearest where the airplane was kept on the doctor's private strip 150 miles distant. Maps were bought covering the 1300 mile return flight, as was an extra-heavy, down-filled coat, for it was in the dead of the northern winter that I would be traversing the Dakotas, Montana, and Idaho during the return flight. All was set. I caught the earliest Friday morning flight possible, intending to be home Sunday evening, weather permitting.

As the flight proceeded east, I began to notice how much desolate, forbidding terrain would have to be crossed on the return trip. This land lay in the icy grip of winter and would be crossed at 500 to 1000 feet above the ground, instead of the 40,000 feet at which the jet now cruised. Plus, the trip would be made at 80 mph instead of 500 mph! A feeling of uncertainty washed through me. Shadows of doubt lingered even after landing, for the airplane I came to buy was an additional 150 miles east of where the airliner stopped.

The owner of the airplane, with whom I had talked a number of times by telephone, graciously agreed to drive the 150 miles from his home to pick me up. He was as nice a person as you could ever ask to meet, and we got along very well from the start. The only problem I felt was that the airplane transaction had taken on the character of a fact—as though the deal was already made, and the only thing left was for me to jump in the machine and fly home. Only I hadn't actually *seen* the airplane yet and was uncertain whether or not it would live up to all the rosy claims that had been made for it. Also, it would be well past dark, yet I was scheduled to leave early the following morning. I would have to make my decision tonight, though I would have no opportunity to really see and fly the airplane. This was turning into a scary proposition, as I could scarcely afford to make a mistake on the purchase.

We arrived at the field where the Champ in question was sitting half in, half out of an open-faced hangar. It was partially covered with frost and snow, and the below-zero temperature of the night caused it to shine back like an apparition in the car lights. I made a brief inspection of the bird, seeing about all I could see in the darkness in a few minutes. The finish was weather checked, and large spots of peeling paint curled up on the upper wing surfaces. As the man had said, the airplane was no showpiece, yet I couldn't help but be a little disappointed at my first glimpse of the plane I was about to lay out my hard-earned and long-saved money for, not to mention that I was about to fly it halfway across the continent in the dead of winter!

I procured the logbooks and asked to be taken to my hotel where I began pouring over the records of the life of this airplane. I had been carefully coached on what to look for by a mechanic friend at home, and almost from page one, there were things in these logbooks that made me uneasy about the purchase.

For one thing, the airframe logs indicated that, for one reason or another, the entire fuselage had been replaced with "an airworthy, used fuselage" several years earlier. To me as a non-mechanic, and in the absence of any elaboration as to the reason for the fuselage substitution, I looked at this as a negative factor for the airplane. Had it been crashed? Had it been in a moist, corrosive environment that had caused deterioration of the fuselage and possibly other structural components? The cryptic log entries failed to shed light on the reason. Further study of the airframe logs indicated that, although the fuselage cover was only about ten years old, the airplane had been painted the same color as was visible beneath the peeling paint on the wings many years earlier. Therefore, the wings probably had not been covered in perhaps fifteen or twenty years. The log entry was written in such a way as to make it very difficult to tell just *what* had been covered and when, or what had not been covered at all.

Moving then into the engine logs for the airplane, I was immediately confronted with the knowledge that, although the engine had been majored, it was the second time it had been done recently. After the first overhaul, the engine had not run properly and had to be torn down a second time, and on top of all this, there was no historical record of the engine's past, as it had been overhauled originally after sudden stoppage during a crash. A *crash*! A crash in what airplane? This one? It was all just too perplexing for me to figure out or understand. Only a skilled mechanic could have really wrung the true meaning out of those logbook entries. I was not a mechanic, but a slightly overzealous airplane enthusiast, hot to buy my first airplane. The owner was as honest as he could be, since he was also not a mechanic and probably did not fully understand the airplane's curiosities and faults any better than I did. He, after all, had not misrepresented the airplane to me, it was just that the picture I had in my eager mind and the one with which I was confronted here did not coincide. And here I was 1300 miles from home—150 miles from the nearest public transportation and with all these doubts. It was a long and sleepless night.

The next morning, after much soul-searching and much flip-flopping between deciding to take the airplane anyway, just to save face, and deciding to decline to take the airplane and wondering how to break the news to the kind doctor who had gone quite out of his way to bring me here. I finally felt that the best thing to do, given the circumstances and my feelings, was to say "no." I did so, and the doctor (disappointed though I'm sure he must have been), seemed to understand my predicament. He even took me all the way back to the airport to catch the airline flight home.

So the point of all this is, be very careful in your long-distance search for an airplane. *Never* fail to do a thorough title search. *Always* scrutinize every page of the aircraft logbooks to determine just what has been the history of the airplane you are thinking of buying. Avoid painting yourself into a corner, verbally or otherwise. Never commit yourself to anything you have not personally checked out. Most important of all, *never* be afraid to say *"no!"*

A postscript to this story is that upon arriving back home in a state of utter disappointment, I wandered out to the local grass airport on Saturday morning, just in time to see a man post a "For Sale" notice on the bulletin board. Upon closer examination, I found the subject of the notice to be a 1947 Aeronca 7AC Champ in good flyable condition for $500 less than the one I travelled halfway across the country to buy! I bought the airplane the same day, after making all the appropriate checks, of course, and have been thrilled with the plane ever since. Things have a way of working out in the end.

Another hint about the choice of and search for your aircraft involves the selection of oddball, rare, or antique aircraft and biplanes. For the antique enthusiast, these airplanes are great. However, for the budget-minded sportplane fan, they can be a financial can of worms. Since this book is designed for the sport flying enthusiasts therefore, I recommend strongly to stay well clear of: (1) any unusual antique aircraft not mentioned herein; (2) aircraft that, though not antique, are rare and generally not readily available on the used aircraft market; (3) biplanes.

The reasons for the first two suggestions deal with the problem of parts availability and cost, while the third caution is primarily due to the much larger increase in expense and the added expertise needed to get involved with any biplane. Biplanes have two additional wings, not to mention assorted flying wires, etc., which can greatly complicate the restoration process. In many cases, they are also harder to fly and maintain and are frequently more expensive to operate. Many have radial engines which are so expensive to maintain and overhaul as to put them well beyond the financial reach of those for whom this book is intended.

Antiques, rare airplanes, and biplanes are wonderful and nostalgic birds, but they are inconsistent with the goal of simple, cheap, fun flying.

Another word of caution I would like to add here has to do with the complete basket case. It is very tempting to buy a totally run-down airplane, one that hasn't been flown in years and has considerably deteriorated sitting out in the weather, because they are often for sale at a very low price—$2000 to $3500. *Buyer beware.* Many times these airplanes are virtually unrestorable or will require so much in replacement parts that the cost of restoration will far exceed the value of the finished product. Some airplanes are so far gone that the average rebuilder couldn't afford them even if they could be had for nothing.

Whether in your backyard or across the country, take time to fully research your prospective purchase.

Chapter 6
Getting Started

For purposes of the restoration project from which the general guidelines and information for the rest of this book is written, I used an Aeronca 7AC Champion, which I bought with all the budget factors in mind and which I feel currently fulfills the equivalent position of the Volkswagen Beetle in the sport aviation economy (Fig. 6-1).

Practical Choice of Fabric Covering

The rebuilder is faced with a variety of fabrics and processes from which to choose. However, there are basically two types of coverings, i.e., those made from natural fibers and those made from synthetics. In the early days, aircraft were covered with real woven cloth. Irish linen was very popular, as was Grade A cotton fabric. These fabrics were easily obtainable, back when most aircraft used them, and they were easy to work with. They were simply cut to fit and stitched over the airframe and wing ribs. Nitrate dopes were used to finish them. Today, linen and Grade A cotton, are still easy to work with, though usually, butyrate dopes are used, and they still give about the best finish available for fabric-covered aircraft. However, there are other problems associated with their use today.

Good quality Irish linen is very hard to find, and to a large extent, so is Grade A cotton. And even when they are available, there are many cases where the quality of the fabric is not up to aircraft standards.

When the rules were originally written pertaining to fabric recovery of aircraft, they were written with these fabrics in mind. It was stipulated that a minimum strength of 80 pounds per square inch be required of cotton when it is brand new. It must be replaced when the strength falls to 56 pounds per square inch. I've recently heard of many cases where *brand new* fabric has come from the mill with a test strength well below the required eighty pounds. It is very important that any new fabric be checked for strength, as it is possible that any given run of fabric may be inferior, but as a general rule, the synthetics such as Ceconite and Stits Polyfiber have a new strength well in excess of 100 pounds per square inch, and remain substantially above the 80 pound strength of new cotton or linen fabrics for many years. Given this basic difference in strength and availability, synthetics are the general rule. Airplanes are being recovered in cotton all the time, however, due to its supposed superior finishing capability over synthetics, and to a lesser extent, due to the long-time preference of genuine fabrics by experienced craftsmen. In actuality, Ceconite, carefully finished, can look every bit as good as cotton or linen.

There are some other considerations in the question of natural fabrics or synthetics. Linen and cotton seem to be more susceptible to the elements, that is water, sun, etc., and so they tend to deteriorate faster when exposed to the elements. On the other hand, however, an aircraft which is well cared for and hangared so as to protect it from the elements will see fabric, no matter whether natural or synthetic, that will last fifteen to twenty years—actually as long as you would want to allow between looks at the remaining airframe parts, wood, metal and control cables.

There is no disputing the fact that cotton and linen finish out better than most of the synthetics. There is also no doubt that synthetics stand up to the rigors of time and the elements better than cotton or linen. The choice is yours. There is one other element, however, you will want to consider in making this decision—that of

Fig. 6-1. The simple lines of the Aeronca 7AC, or any of the immediate post-WWII lightplanes for that matter, make restoration relatively easy and inexpensive, and they are lots of fun to fly!

cost. The cost of good quality Grade A cotton is now far in excess of the cost of Ceconite, or practically any other synthetic aircraft covering material. All things considered, my choice for recovering a sport airplane is synthetic fabric, for economy, durability and ease of workmanship.

In the upcoming description of how an airplane is recovered, I will use Ceconite as an example. Ceconite is one of the more popular synthetic fabrics, and there is relatively little variation in the techniques or procedures used in recovering an aircraft, no matter whether you use natural or synthetic fabric.

Work Area Considerations

A very big consideration, and very often one of the hardest requirements to satisfy, is the selection of an adequate work area for your restoration project. There have been cases where dedicated rebuilders have used basements, living rooms, kitchens, extra bedrooms, etc., where a workship was not available. Let me assure you that the project upon which you are preparing to embark will be made much easier if you can obtain an adequately large, well-ventilated utility area that is equipped with sufficient electrical outlets, and can withstand the spraying of dopes, paints and the dust storms of sanding and sand blasting. Most homes are not adequate, nor will you want to attempt to live in the disarray of a three-month or longer recovering job. A garage is the very minimum work area, actually, and a large one-car, or average two-car garage is quite satisfactory.

You will need room to move large items in an out, such as the fuselage and wings; you will need enough room to turn these items around so that you can work on them, spray dope and paint, etc., and you must be able to move about your work on all sides, so sufficient work space is tantamount to doing a good and thorough job.

Do not attempt to recover an airplane in a garage in which you are also keeping your car. Invariably, overspray from the doping and painting process will find its way onto the finish of the car. Beware of leaving anything that would not look good painted the same color as your airplane in the garage where you are working.

Also, if you decide to use a garage, check your homeowner's insurance policy to make sure you are not violating the terms of the policy by doing this type of work, which includes the spraying of flammable substances. Although you will have a fire extinguisher at hand at all times while working with the flammable dopes and paints, it is essential that you don't take on the greater risk of nullifying your fire insurance protection as you work on your project.

The type of heat source used in the work area should be checked also. If the heat source is electric, for example, with heat elements exposed to the air, you must be extremely careful to shut down the heaters prior to spraying dope or paint, in order to reduce the fire hazard. Always keep in mind that while these dopes and paints cease to be highly flammable once they are dry on the aircraft surface, while they are still wet and in the form of spray, they can and will ignite rapidly and burn fiercely.

In your search for a work place, ask around at the airport. You may discover that the local FBO has available a small room in his shop, or in a hangar on the airport, which is not only designed for the type of work you're doing, but puts you in closer proximity to materials and expert advice. Once when I was starting a rebuilding project, I discovered an entire shop in an abandoned hangar on our municipal airport, complete with hoists, work stands, lights, ventilation and forced air heat—all for $15 per month. I found the shop by accident after explaining to acquaintances that I was experiencing difficulty in finding a place to do the work. It pays to shop around.

You will undoubtedly find, if you haven't already, that if you are having any kind of trouble related to aircraft, it is always a good idea to be very vocal in expressing your difficulty around an airport. Most times, help will be offered immediately, since airport regulars always know the best (and cheapest) way to accomplish practically anything in aviation!

One other word about using old shops and the like around airports: ask the airport fire marshall, if the airport has one, what the rules and regulations are with regard to spraying dopes and paints inside the building. In some cases, this is prohibited, and in others, certain requirements will prevail for the use of fire extinguishers, etc. Stiff fines and the interruption of your project could result from inattention to the fire regulations.

Just remember that you do not have to have a place big enough to house the entire assembled airplane; you will be removing the wings and all control surfaces, and you'll only be working on one part of the ship at a time, so as long as you have enough room to maneuver around the fuselage with the wings off, this is about the biggest piece of the airplane that you'll be dealing with at any one time.

Be sure that the work area has enough electrical outlets. You will be using an electric drill, an air compressor, an iron, and several lights. Although you won't be using these things all at once, a minimum of two to four electrical outlets is a must.

You must also be able to maintain a temperature of at least 72° F during the time you are spraying dope and paint, and either humidity

Fig. 6-2. Inside a hangar is the best place to start dismantling your project airplane.

must be low, or you must use a dehumidifier. While the temperature must be kept relatively high, enough ventilation must be maintained to keep the air cleared out from the heavy fog of dope or paint spray. Enough windows or doors must be capable of being opened while you are spraying. Never try to spray a coat of paint or dope without this ventilation, or you may immediately float to the ceiling and forget why you are even there.

Disassembly

Now that you have made the choice and have purchased your airplane, let's assume that you have purchased it with the intention of going right ahead with the rebuilding project. Of course, it is entirely possible that you found an aircraft that can be flown for a few months or years before reaching the point of needing recovered, but sooner or later the job has to be done. Let's get started.

Whether you do the work at home or at the airport, you will be starting with the aircraft at the airport and disassembling it there, at least as far as removing the wings and tail surfaces. If at all possible, avoid tearing into the airplane outdoors. Put it in a hangar, if you can, at least until you have it far enough apart to either put it into your airport shop space, or truck it home to your garage. Even on calm, clear days, other airplanes, onlookers, the wind and the odd rain squall can really upset the careful work that you will be doing (Fig. 6-2).

Once the aircraft is sitting in the hangar where it can be left unmoved for a couple of hours, get at least one helper, and begin removing all fairings, inspection plates, and moldings (Fig. 6-3). Draw the locations of the inspection rings on the aircraft diagram so that you can remember where they go during the project (Fig. 6-4). You will want to do the same for all control cables, and anything that has to protrude through the fabric surface, since holes must be cut at precise locations.

At this point, you can check the service manual for the airplane, if you have one. If you do not, you can usually find a service manual for your airplane through Wag Aero, a large supplier of service and other information on many out-of-production airplanes. (See Sources of Materials and Supplies for address and phone number.) This will have diagrams that tell you where all control cables, attachments, brackets and fittings are, as well as the numbers of all bolts and hardware, should you need to replace them. For examples of the illustrations, diagrams, and other descriptive data to be found in an aircraft service manual, see Figs. 6-5 through 6-16.

Once the fairings and inspection plates have been removed, disconnect all control cables for the ailerons, elevator, elevator trim and rudder. Some, such as the aileron cables and flap control rods, if so equipped, will have to be disconnected from inside the cockpit, so that the cables, etc. can come away from the fuselage along with the

Fig. 6-3. Start by removing all fairings and the cover attachment points for wings and/or control surfaces.

55

Fig. 6-4. Remove all inspection plates, but mark their location on a sketch of the airplane for reference when the time comes to replace them.

wing. Other cables, such as the tail control group, are best disconnected right at the controls, so that the controls can be removed from the fuselage. Make notes in your notebook as to where each control attachment goes, and record other data, such as the method of attachment, bolts, screws, cotter pins, and the like, and label, as clearly as possible, each part removed (Fig. 6-17). This may seem like a lot of unnecessary work now, but you will greatly appreciate having done this when the time comes to reassemble the airplane after you have had time to forget where everything goes and how it is attached to the airplane. Again, service manual diagrams will help.

Once you have unfastened all of the control cable connections, take all the cotter pins and safeties off the strut attachment fittings and the wing root attachment fittings where the wings attach to the fuselage. Loosen, but do not completely remove the nuts on these attachments (Fig. 6-18).

Now get up inside the wing root between the wing and the fuselage and carefully look at the situation. You will want to make certain that all fuel lines, instrument lines, tubes and wires are disconnected between wing and fuselage. You will soon be removing the struts and the wings themselves, and you will not want anything hanging on to the fuselage that will preclude the removal of the wing.

Prior to actually removing the struts and wings, arrange to either attach an overhead hoist to a sling beneath the wing surface, or have one or more helpers available to support the wing once the struts are undone and the wing is being removed. Two people can accomplish the entire removal of the wings, but three or four are better. It is desirable, though not mandatory that your mechanic be involved in this dismantling process.

When you are sure that the only remaining attachment of the wings to the fuselage are the wing attachment bolts and the strut attachment bolts, you are ready to disconnect the struts. Be sure to have saw horses available, or a suitable place to put each wing before starting on the first one. Once you and your helper start removing wing attachments, you are pretty much committed to finishing the job and there will be no opportunity to stop to reposition your equipment. If you are using a ceiling hoist and sling, this is all the much easier, but usually these items are unavailable and simple brute manpower is used to support the wing as it is disconnected.

Remove the strut bolts from the wing, then from the strut/fuselage attachment bracket. From this point on, someone or something will have to support the wing near the tip at all times until you have a chance to remove the wing/fuselage attach bolts. Once they are removed, simply slide the wing attach bracket out of the fuselage attach bracket and the wing is off. Have no fear that the airplane will tip over with one wing on and the other off. I had this same apprehension during my first project, and was greatly relieved to see that the airplane is perfectly content to stand upright, even in this very lopsided condition—at least the Aeronca 7AC is, anyway.

Conduct the same procedure on the other wing, and then start on the tail group, disconnecting and labeling each brace wire and attachment point (Fig. 6-19). Save all nuts, bolts, washers, screws and bushings (Fig. 6-20). Though you may want to replace some or

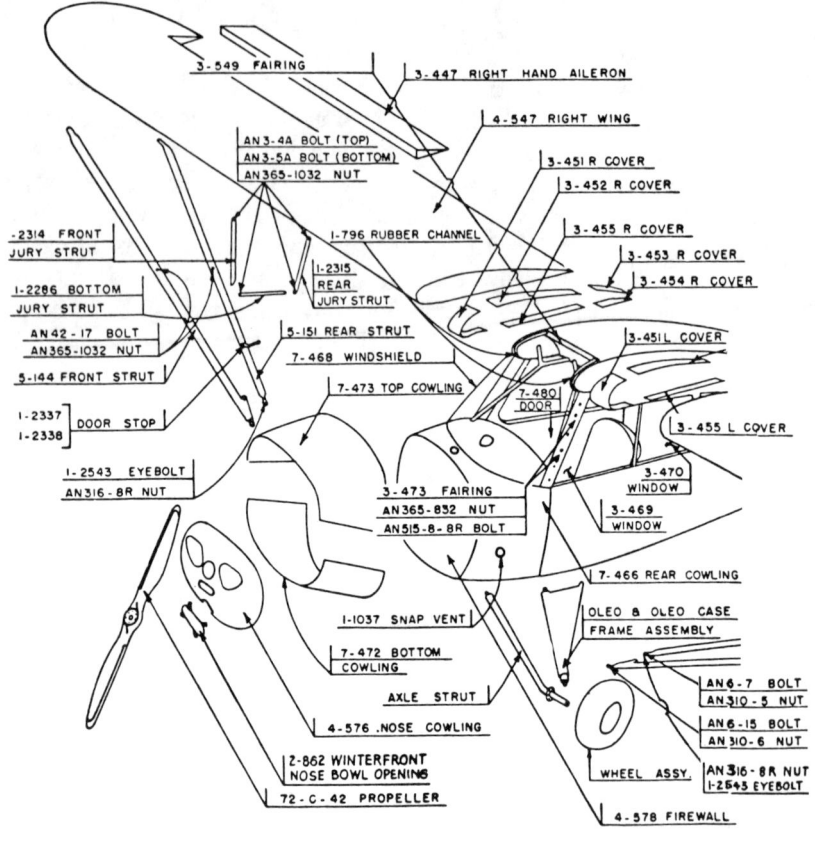

Fig. 6-5. Service manuals are usually either original factory manuals, or reprints, such as these. This is an example of one of the illustrations found in an Aeronca 7AC service manual. Be sure to consult the appropriate manual for your airplane. Each airplane is different (courtesy Wag Aero).

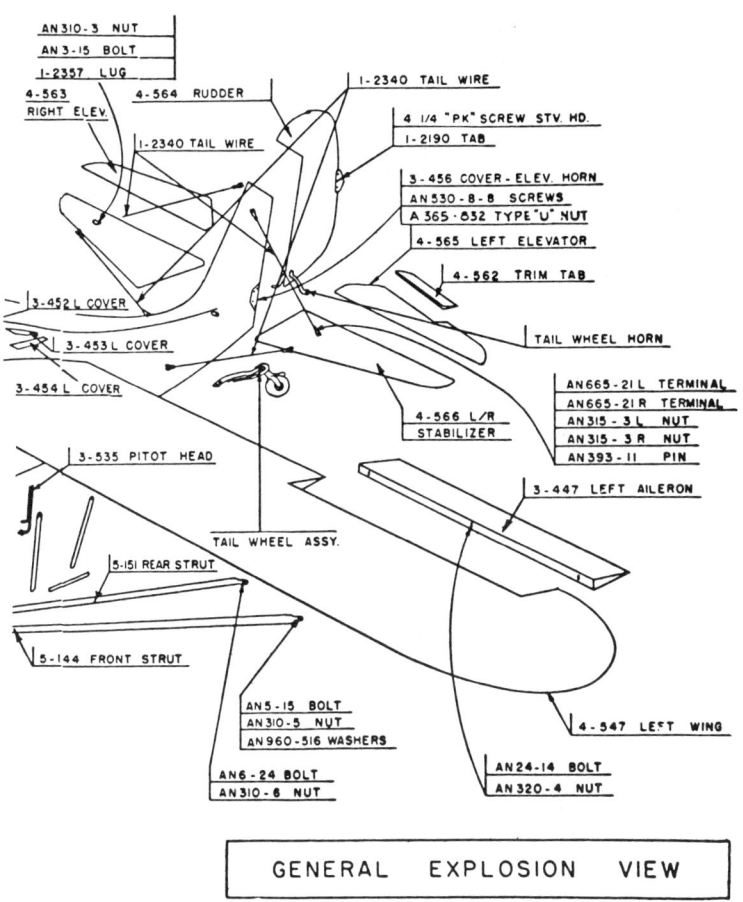

GENERAL EXPLOSION VIEW

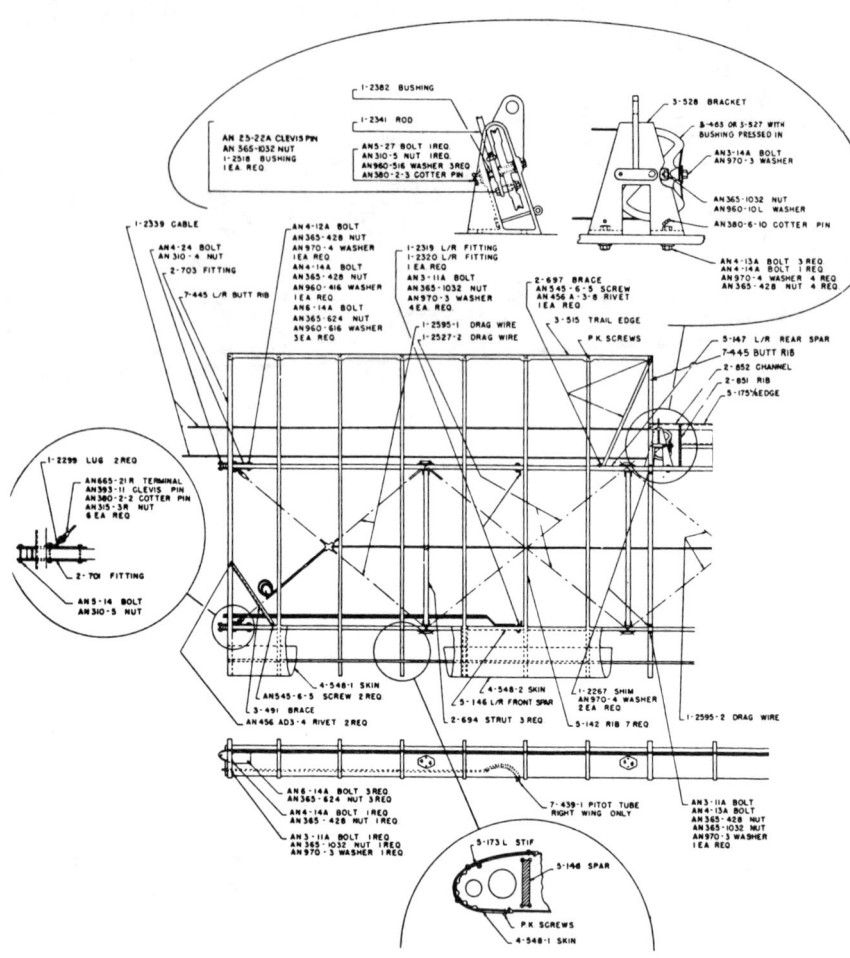

Fig. 6-6. Specifications for all components and hardware can be found if the service manual for your airplane has diagrams such as this one detailing the wing frame assembly of the Aeronca 7AC Champion (courtesy Wag Aero).

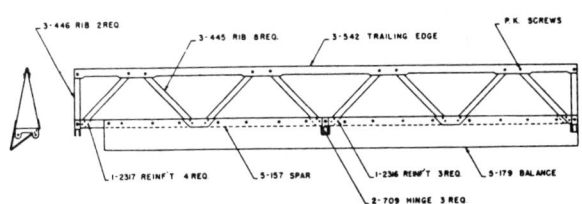
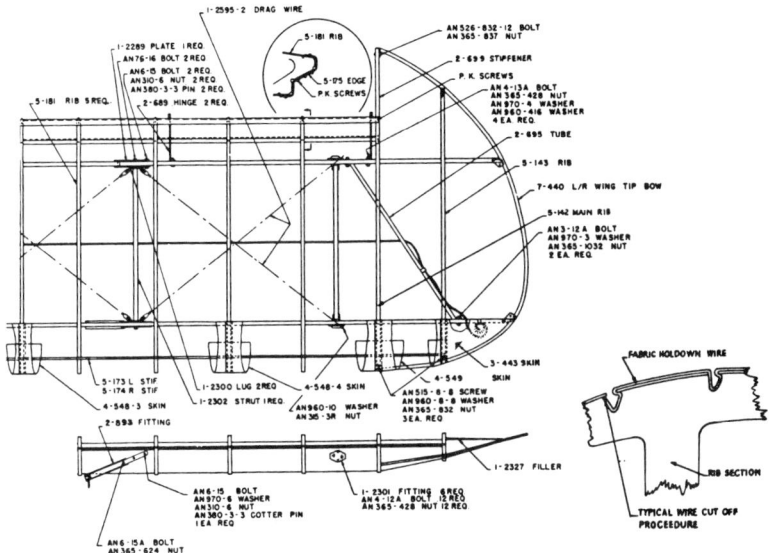

7-439 L/R WING FRAME UNCOVERED
5-180 L/R AILERON FRAME UNCOVERED

WING FRAME ASSEMBLY

all of them, depending on their condition, it is extremely helpful to be able to account for each piece of hardware when ordering new parts or reassembling the aircraft. Once the wings and tail group are off the airplane, you can then place the aircraft in your airport workshop, or trailer it home to a shop or garage (Fig. 6-21).

Once situated in the shop, it is time to begin removing the old fabric and then preparing the airframe itself for the recovery.

Set up the shop to accommodate working on the fuselage first, and then the wings. The reason for this sequence is that the fuselage has to be taken out for sandblasting, etc., so you can begin working on the wings at that point (Fig. 6-22).

Using a carpet knife or other heavy utility cutter, split the fabric down the back of the fuselage from just aft of the windshield to the base of the horizontal stabilizer. Do the same thing underneath the airplane, from the firewall behind the engine all the way back to the end of the tail. You are now ready to peel the old fabric completely off the fuselage.

Begin at the top of the cabin, behind the windshield, and start pulling the fabric back, one half at a time. Use the knife to cut as you go. In all probability, the fabric will tear off with minimal resistance, particularly old, rotten fabric. Work your way back to the tail, pulling the fabric loose all the way around one half of the fuselage frame. Do the same with the other side (Figs. 6-23 through 6-26). In a matter of ten or fifteen minutes, you will practically have stripped the entire fuselage of fabric, except for perhaps the parts of the vertical tail surface that have stitched or screwed-down retainers to the underlying structural members. The Aeronca has the screw, and these along with the cloth buffer tapes must be removed before the fabric itself can be peeled off (Fig. 6-27).

Once the fabric has been completely removed from the airframe, roll up the assorted pieces of old fabric and store them aside for later reference, in case you need to determine the location of inspection plates, control cable openings, drain grommet locations, etc. (Fig. 6-28).

Once the fabric is all removed, there is a considerable amount of work involved in removing the interior of the airplane, including the headliner, seats, carpets and controls. The instrument panel will come out too, as will all windows, the windshield and anything else that is fixed in the cabin of the airplane (Figs. 6-29 through 6-31).

To do a complete job of preparing the fuselage, removal of everything aft of the prop spinner is required.

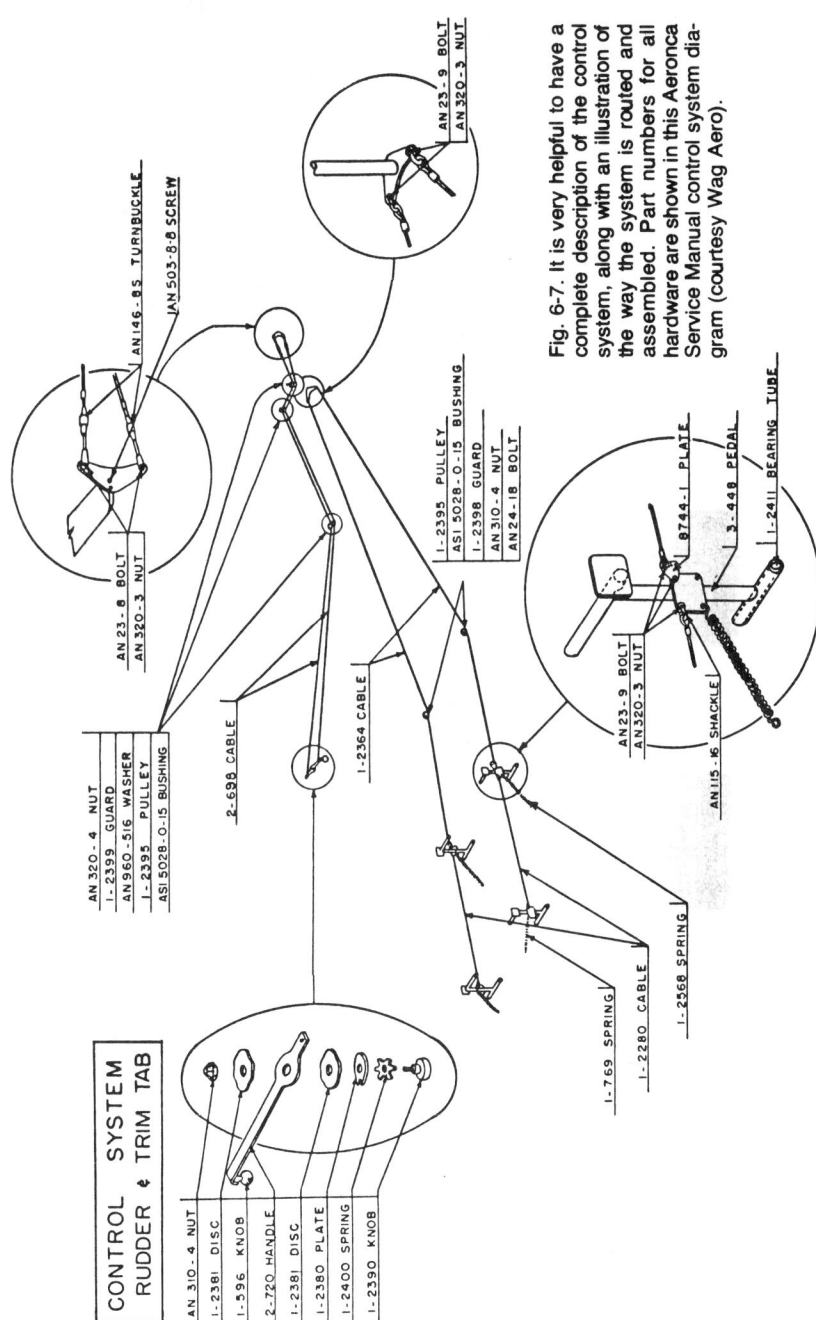

Fig. 6-7. It is very helpful to have a complete description of the control system, along with an illustration of the way the system is routed and assembled. Part numbers for all hardware are shown in this Aeronca Service Manual control system diagram (courtesy Wag Aero).

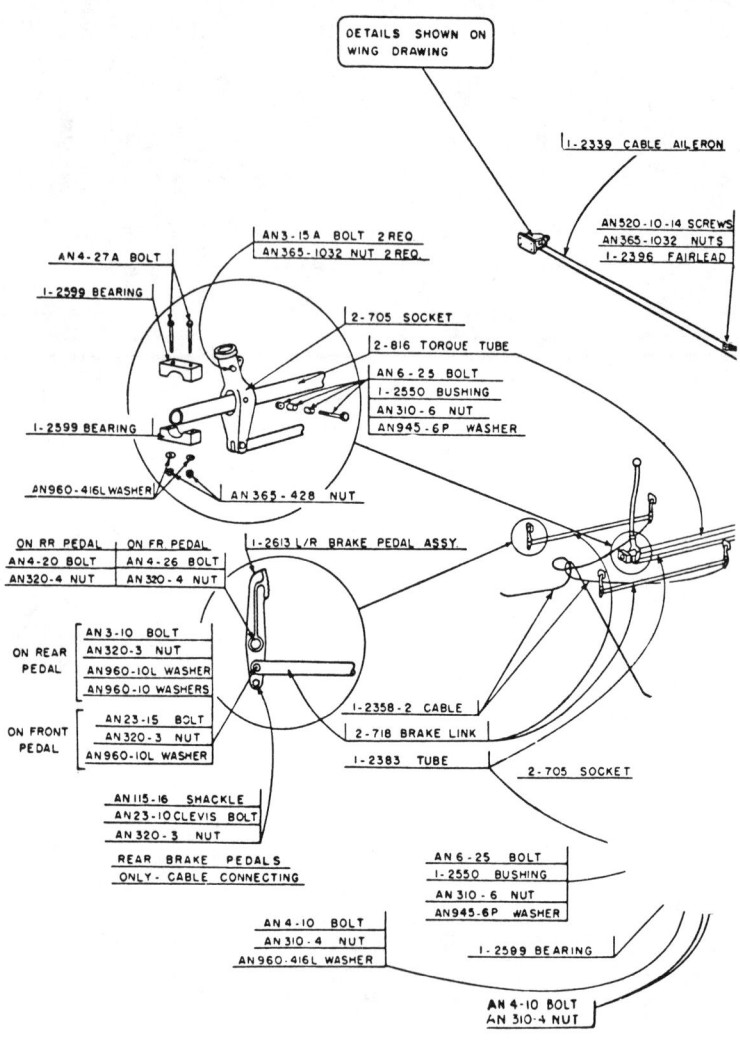

Fig. 6-8. Control system: aileron, elevator and brake (courtesy Wag Aero).

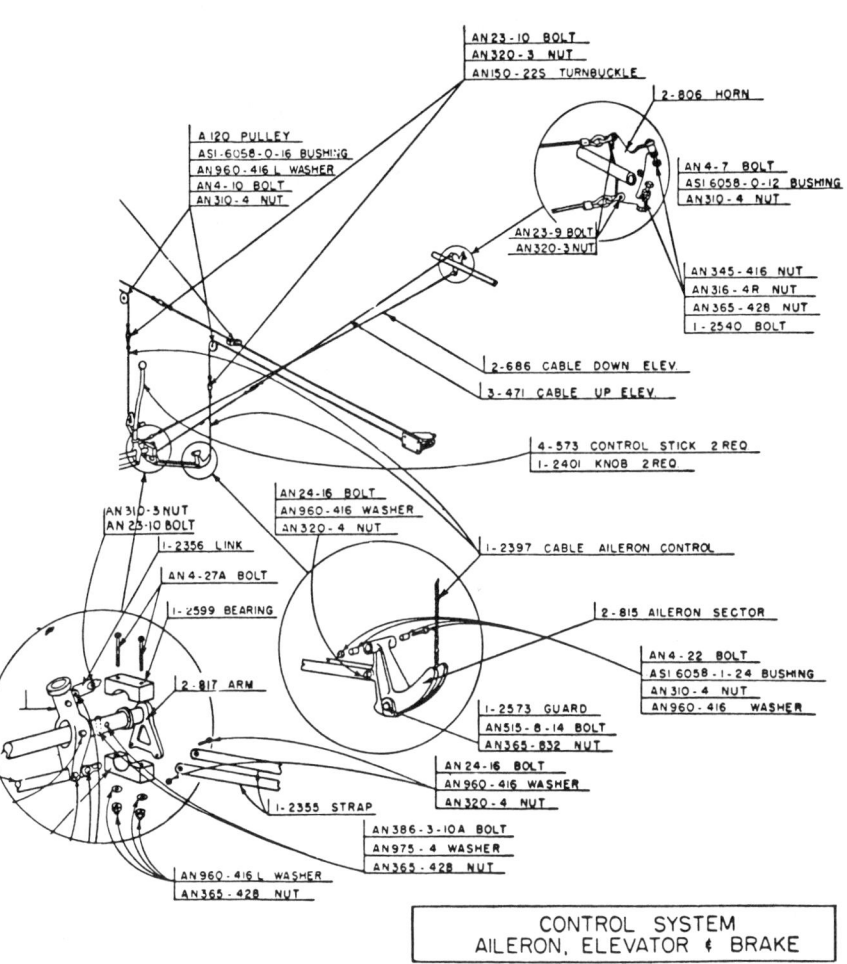

Take off the propeller, all cowling and baffling for the engine. Then unfasten all electrical and fluid lines, along with all controls for the engine: carburetor heat, throttle, etc.

An overhead hoist is a practical necessity for the removal of the engine, as is the help of a qualified mechanic. Once the engine is free of the airframe, the firewall will be easy to remove. Next comes the fuel tank, if your airplane has a fuselage-mounted nose tank like the one in the Aeronca 7AC (Figs. 6-32 through 6-34).

At last you are down to the main airframe, with little left to go before sanding or blasting can be done.

Now, you are ready to begin a very thorough inspection of the wood airframe components and the metal airframe tubing.

Inspecting the Airframe

Once you have removed all the fabric, you will then be able to get a good look at the condition of the airframe, the wood structural members and metal tubing. You must be very thorough in your inspection for wood rot and for rust and corrosion of the metal airframe. Your mechanic will have to look at these components as well, in order to advise you as to whether you will need to replace all or part of the wood, and whether or not repairs will be necessary to the metal tubing (Fig. 6-35 through 6-37).

As I have stated earlier, I strongly recommend that if the wood formers of your aircraft are questionable, they be replaced with a new wood kit, if one is available for your airplane, or that new wood is put in place of the old by hand crafting the members from aircraft quality wood. This is because you are about to recover the airplane and make this wood inaccessible for another fifteen-odd years. This is your last chance for that long to do something about bad wood, and the price of replacing wood with a new kit is not excessive. I emphasize that I am dealing with wood members that can be purchased in kit form, such as was the case with the Aeronca 7AC, because it is beyond the scope of this book to give a detailed description of the methods of constructing structural members for various aircraft. This is an exacting process that must be accomplished by skilled aircraft mechanics that have had the experience (a rare breed). On the other hand, wood kits are pre-cut and pre-formed so that all the rebuilder has to do is simply attach the members according to the instructions, in a manner exactly the same as the wood that was originally there. This is where a thorough diagram of the airframe will come in handy in helping you during the reassembly. For most light aircraft, a service manual can

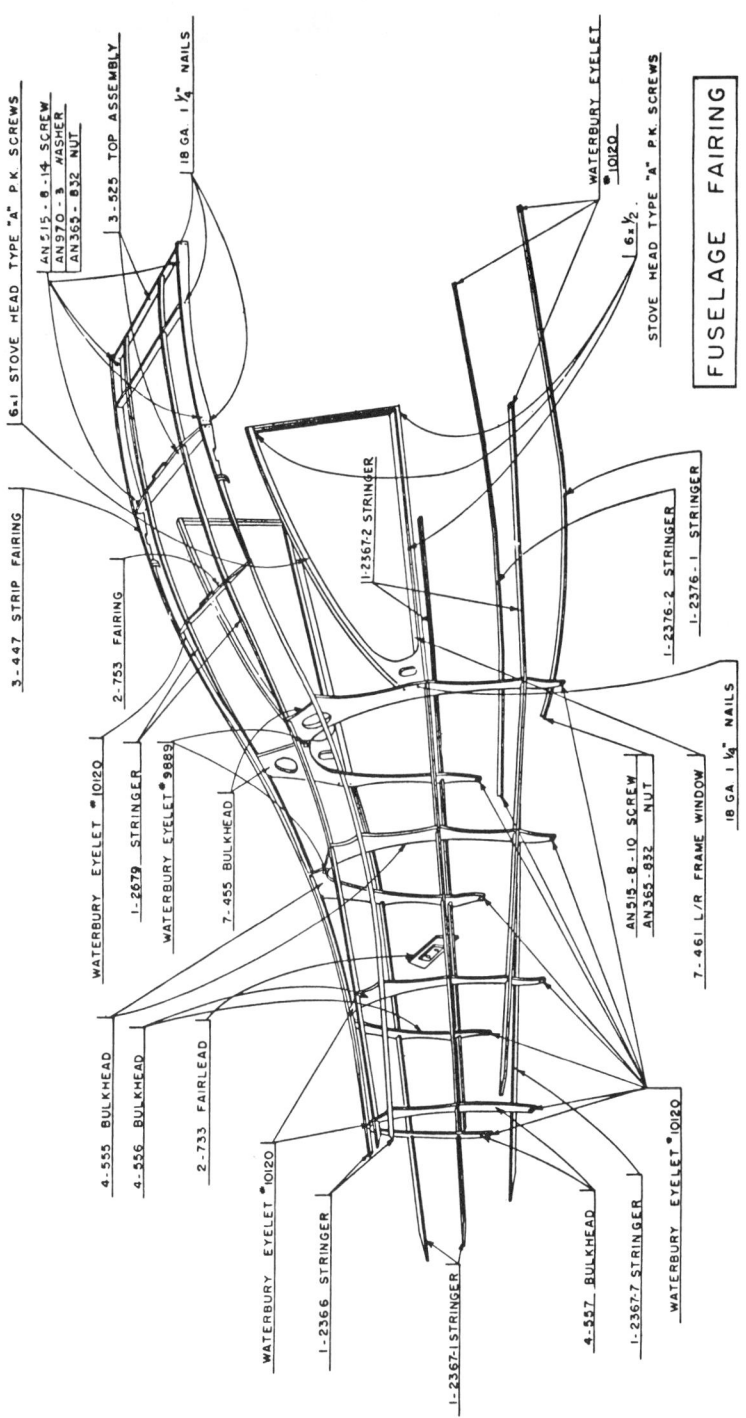

Fig. 6-9. Here is an excellent diagram of the wood fairings that make up a portion of the form and structure of a fuselage. Not all airplanes use wood fairings, but most of the older, fabric-covered airplanes use wood to some degree (courtesy Wag Aero).

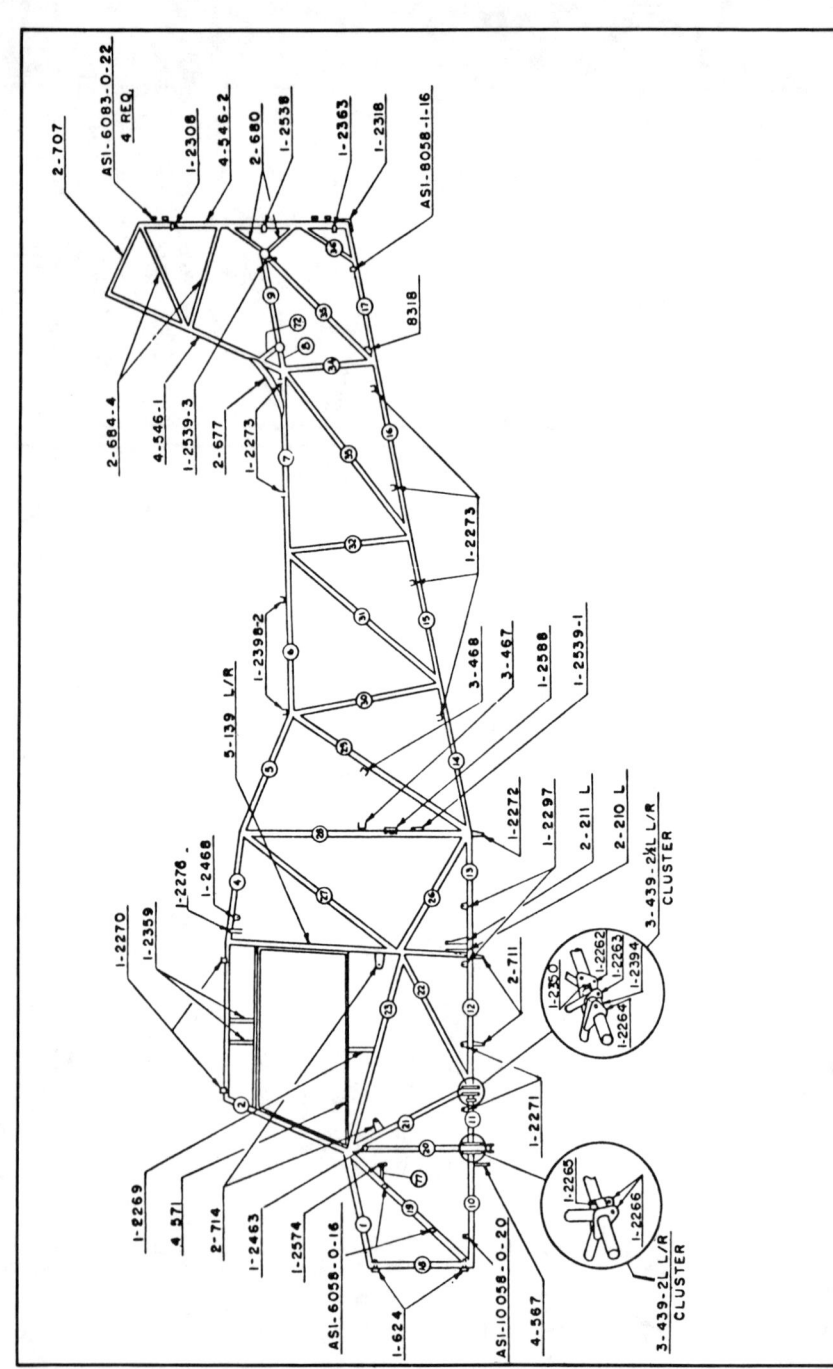

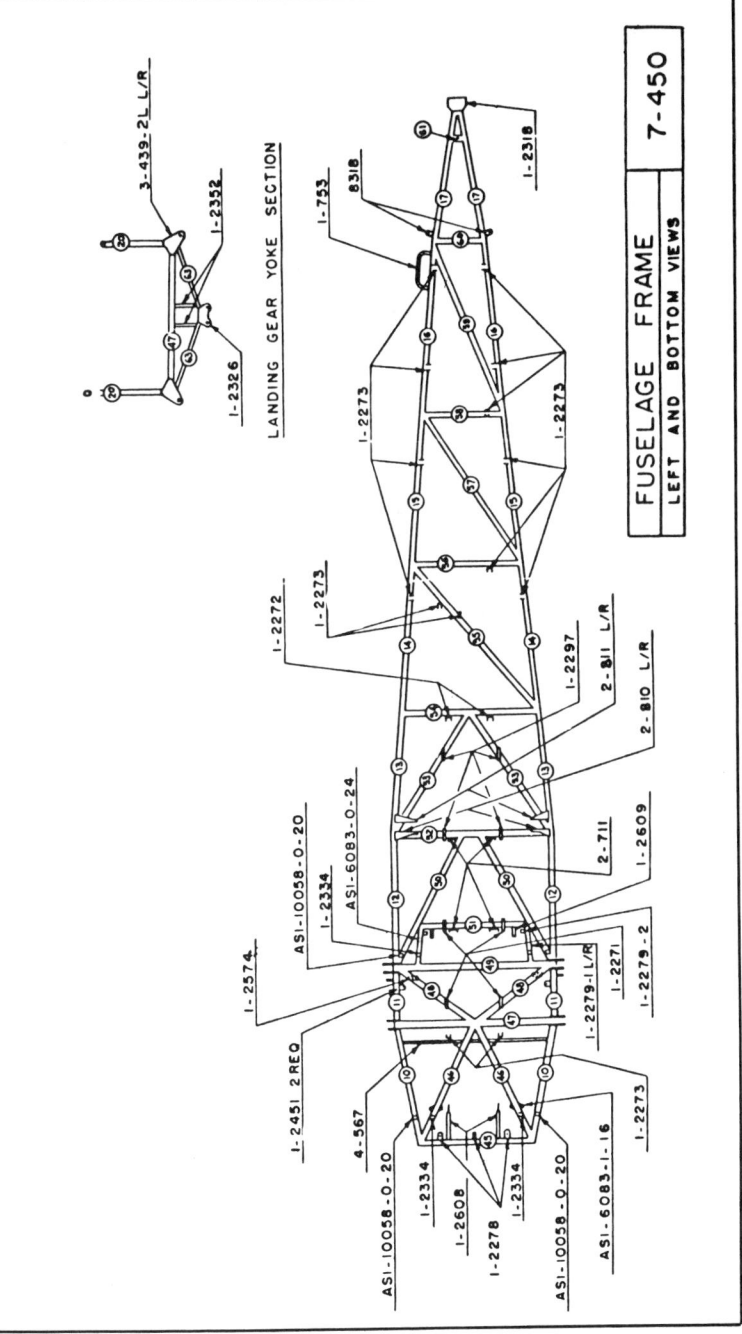

Fig. 6-10. Any questions about the configuration or hardware used in the metal fuselage frame can be answered by consulting the diagrams in the service manual. Part numbers and even the sizes of screws, bolts and nuts are given on most service manuals of this type (courtesy Wag Aero).

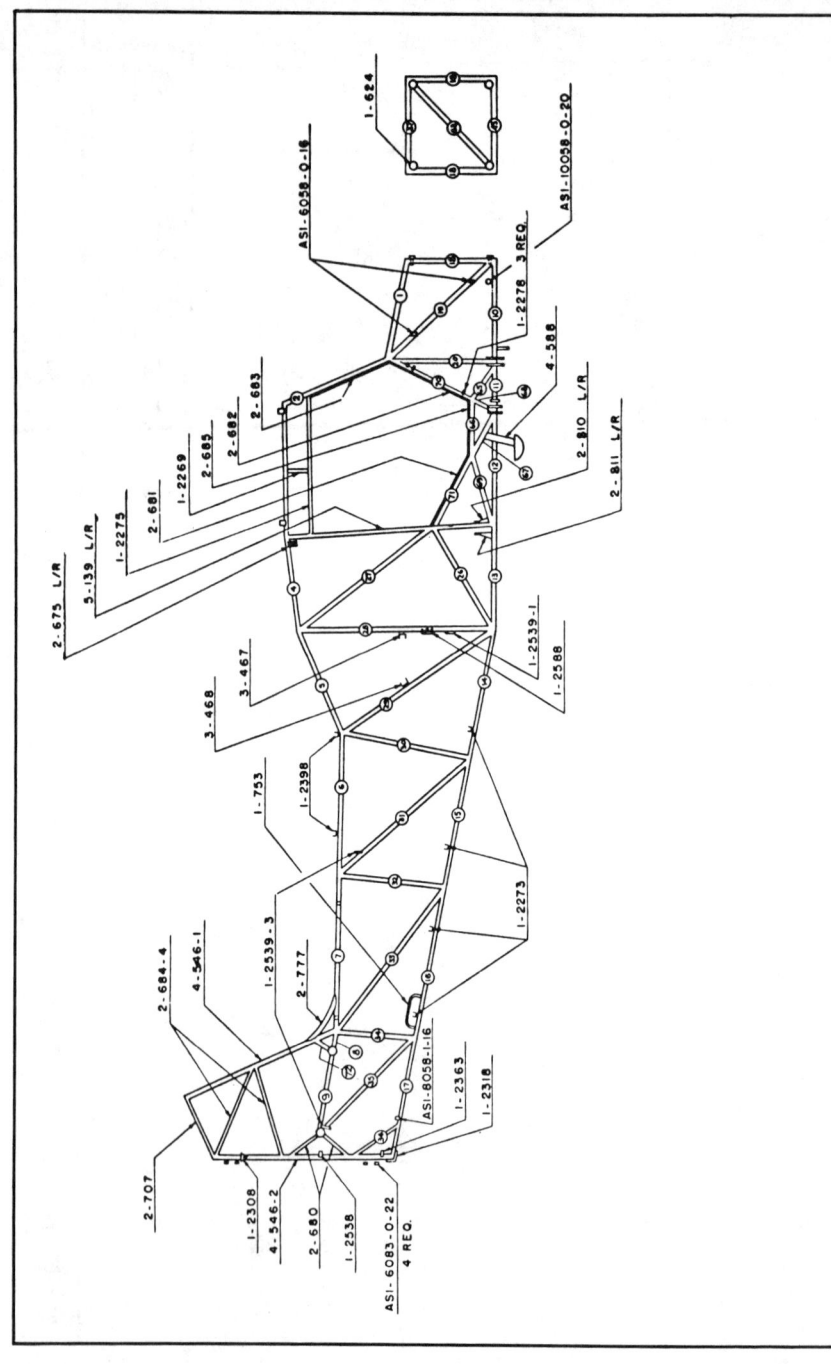

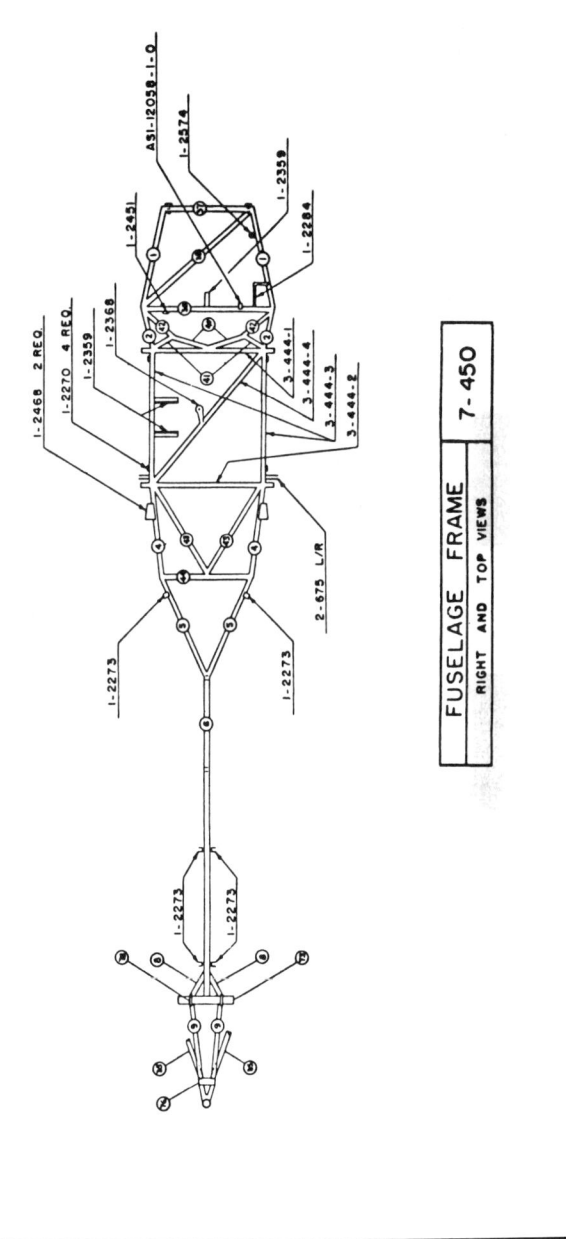

Fig. 6-11. Fuselage frame, right and top views (courtesy Wag Aero).

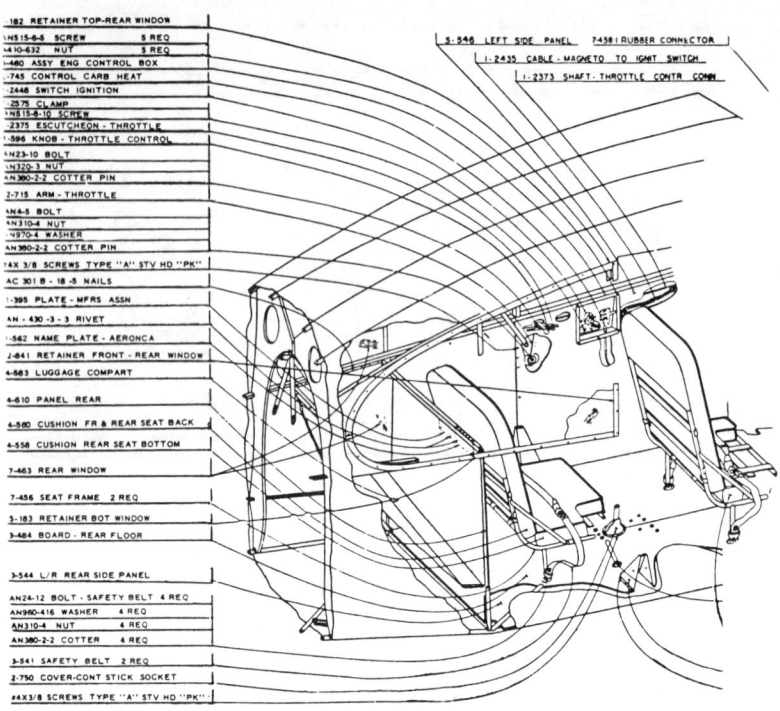

Fig. 6-12. A fixed equipment diagram will tell you where things are located in the cabin (courtesy Wag Aero).

be found that will give a detailed description of the method of attachment of the wood, and will even give the numbers of the appropriate screws, bolts and other hardware and the location of attachments. Since aircraft vary in the amount of wood used in the fuselage and the method of attachment, I will leave this to you, your ability to sketch out the way the old wood was attached, and the advice you will seek and use from your mechanic/supervisor (Fig. 6-38).

Once the decision has been made as to what to do with the wood, you will have to decide what the metal parts of the airplane need in the way of refurbishing.

In most cases, some rust and/or corrosion will be evident when you open up the fuselage. Peeled primer paint, and blotches of rust like that shown (Fig. 6-39), will very likely meet your nervous gaze as the light of day strikes the long-hidden tubing. But fear not, (at least at first) for this is to be expected. It is the *degree* to which rust and/or corrosion has attacked the metal that is in question. An aircraft that has not been recovered in the last fifteen to twenty

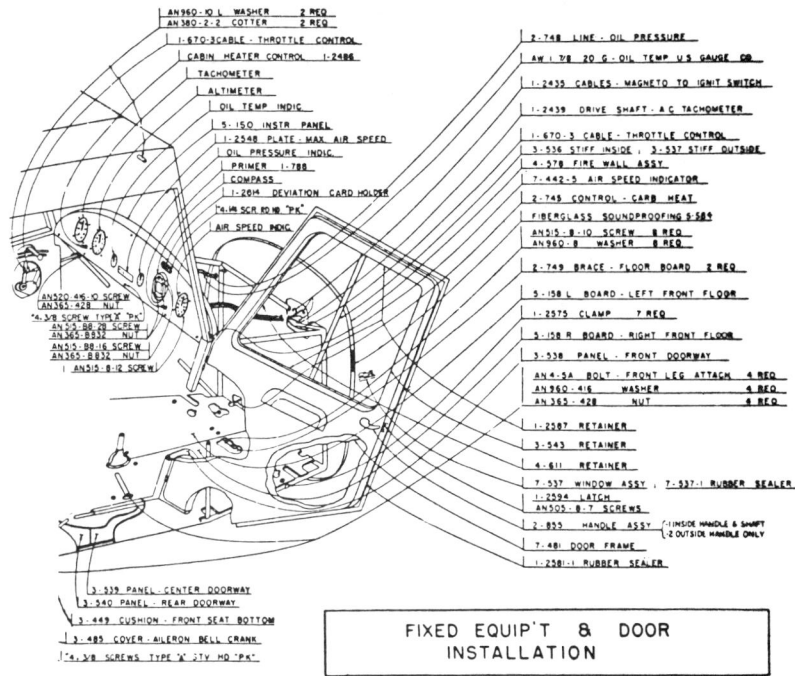

FIXED EQUIP'T & DOOR INSTALLATION

years, especially those that have been sitting outside in the elements, will probably have rust and it is up to you and your mechanic to decide if the airframe is, or is not, repairable. It is rare that a carefully-chosen aircraft that is not an obvious basket case would harbor an unrepairable airframe. Some cutting and splicing may or may not be necessary.

If a fair amount of rust exists, I strongly recommend a good professional sandblasting job be done on the bare metal airframe. To do this, you will have to strip the airframe down to absolute skeletal remains, and this means the removal of all wood, cabin interior, controls, seats, floorboards, instrument panel, fuel tanks, wires, the engine and firewall—every single piece that is attached to that airframe must come off. It is a big task, but one that is worthwhile to restore your aircraft to mint condition. It will be worth far more for having done this. If you do this, carefully label each control cable, handle, instrument, bracket, etc. for easier reference during reassembly (Fig. 6-40).

A good way to determine how deep rust has penetrated metal is to begin with a stiff wire brush and see if you can knock off some of

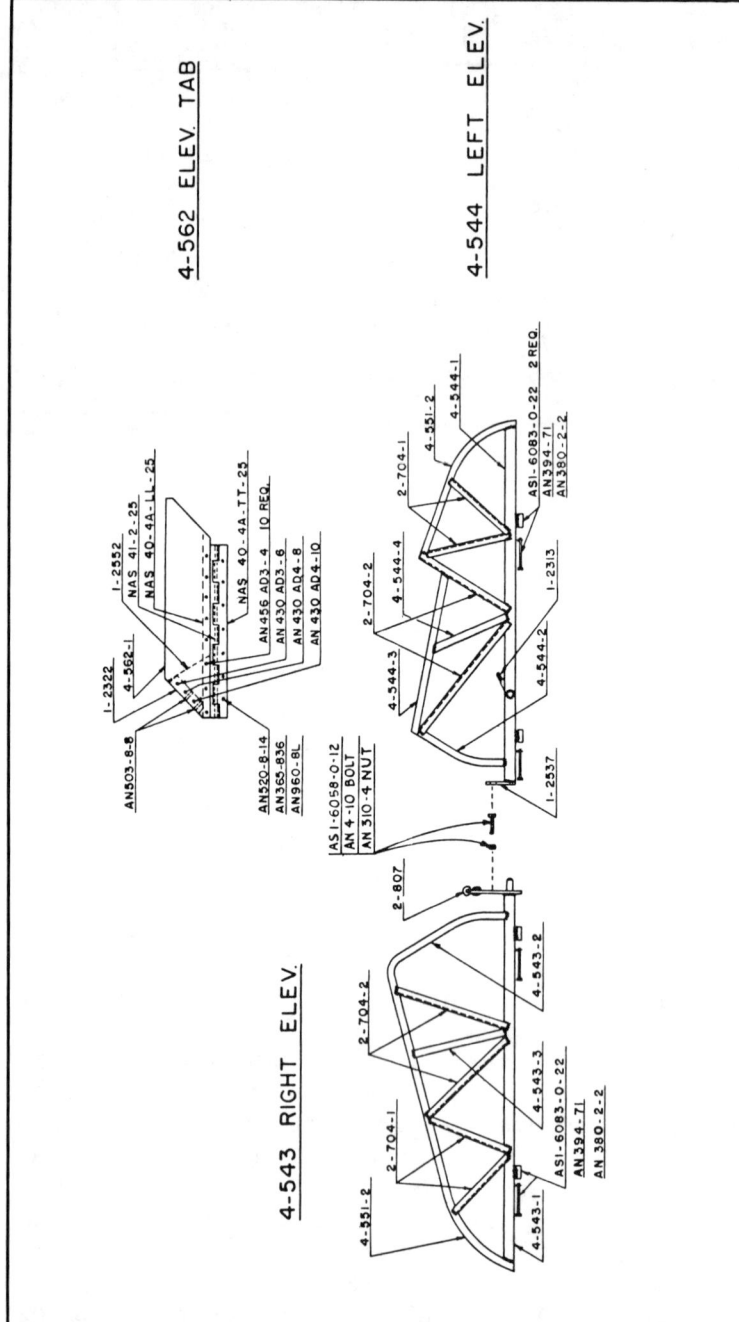

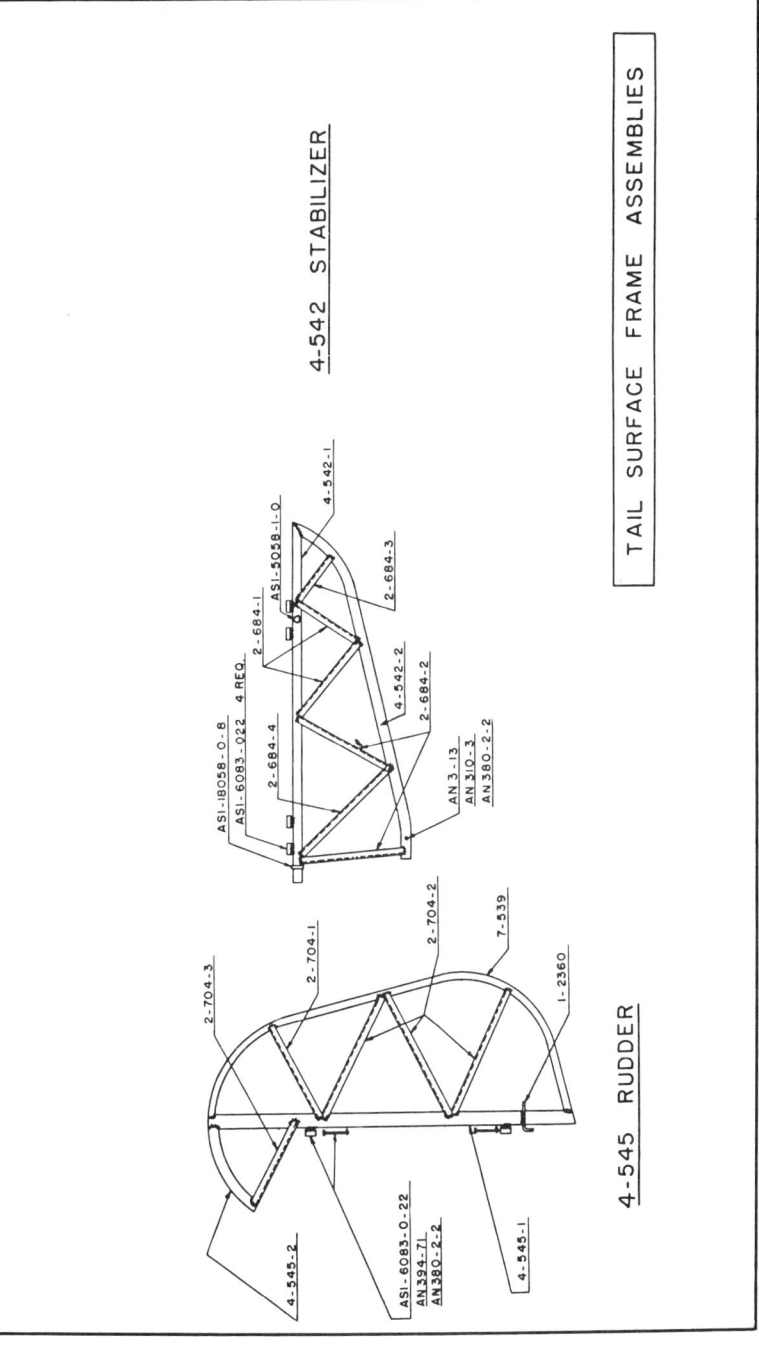

Fig. 6-13. Control surfaces are also described in detail to enable easy repair and reassembly (courtesy Wag Aero).

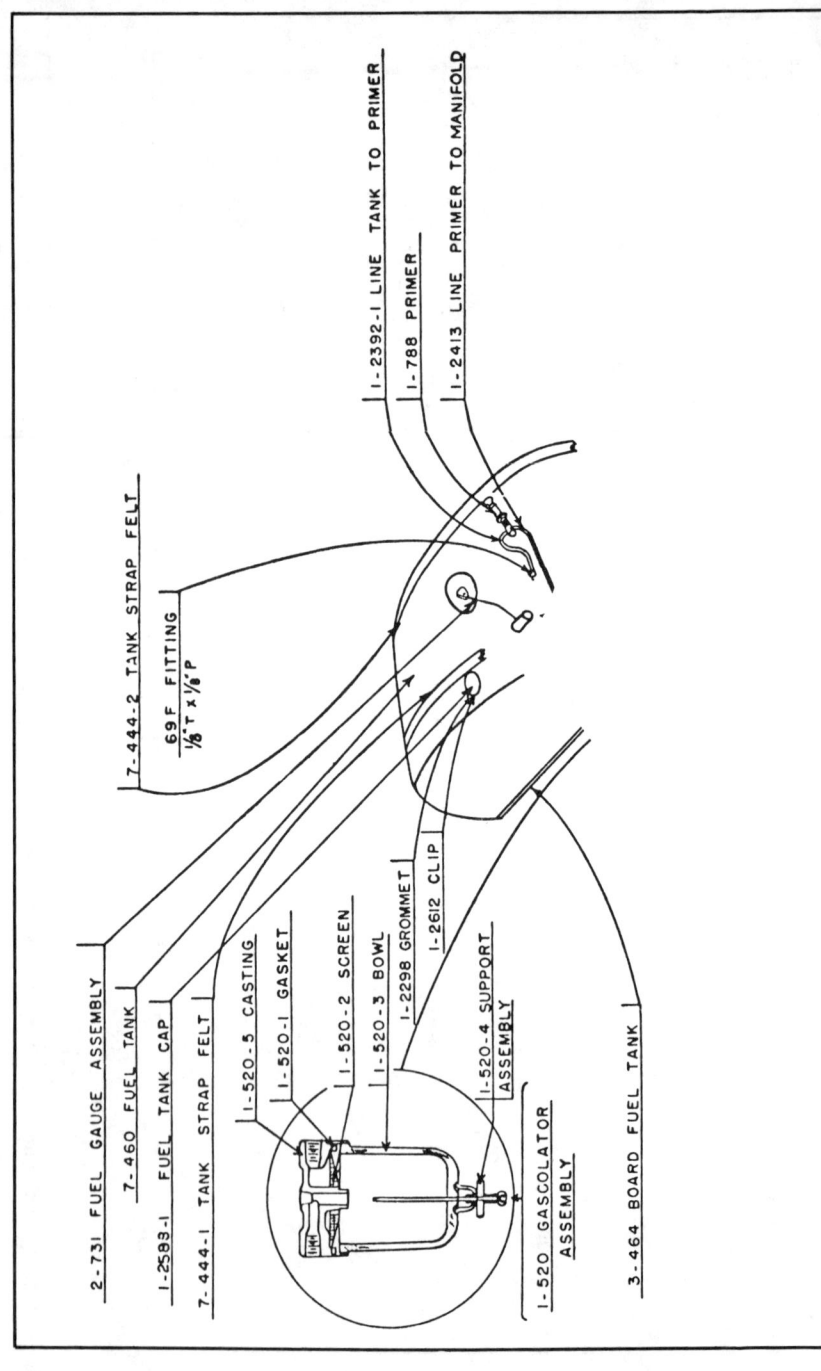

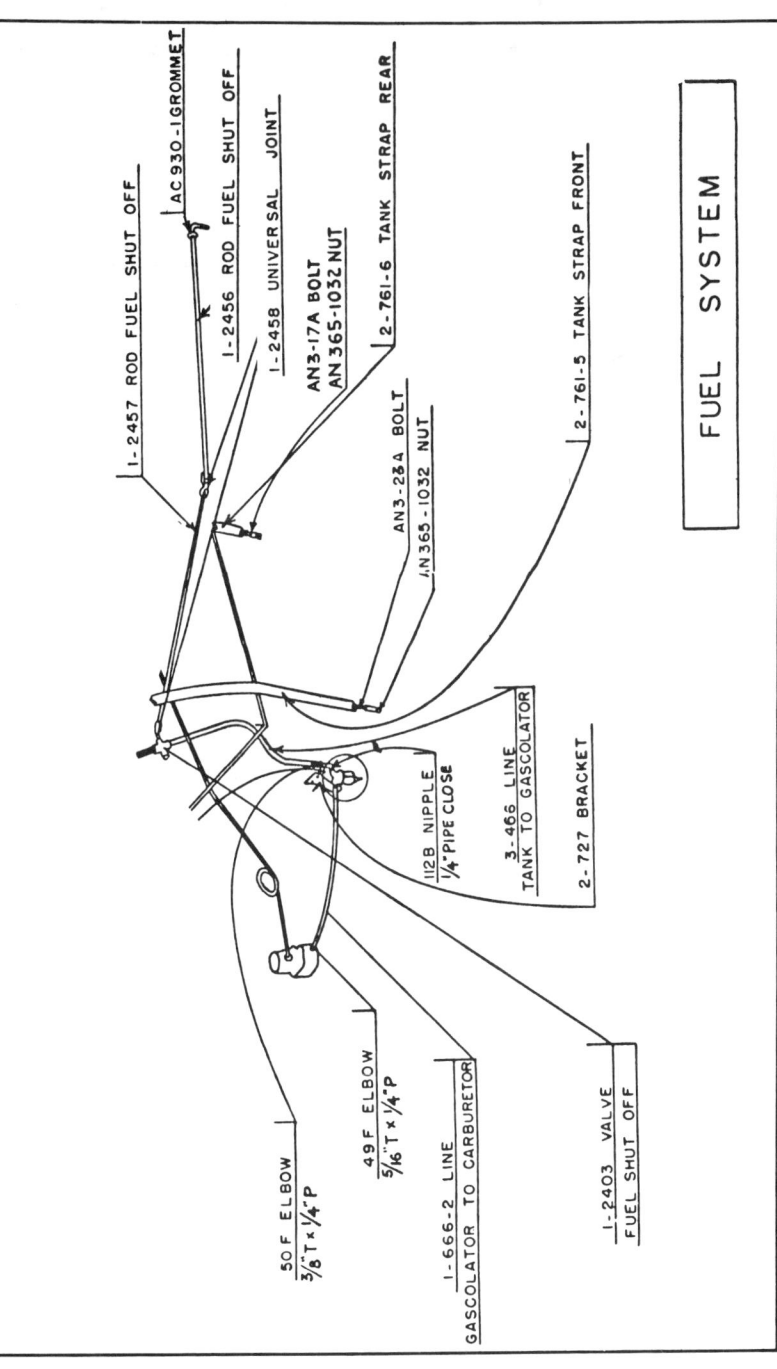

Fig. 6-14. This diagram details the fuel system of the Aeronca 7AC (courtesy Wag Aero).

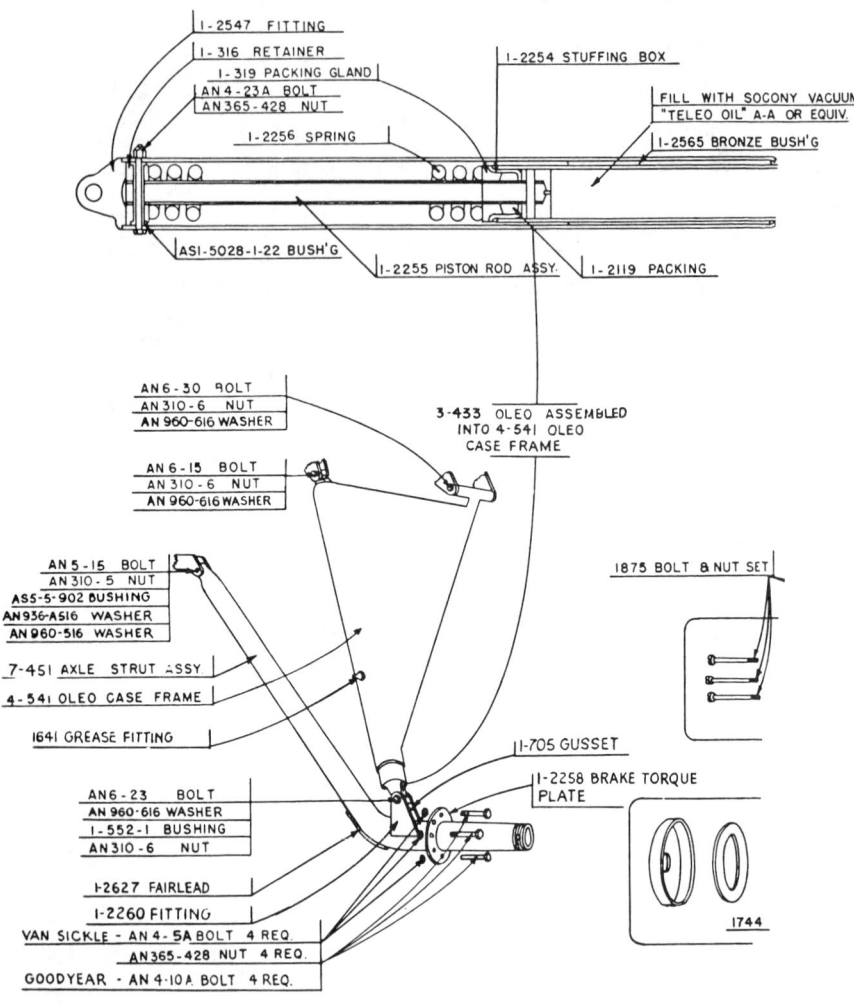

Fig. 6-15. Main landing gear and tail wheel (courtesy Wag Aero).

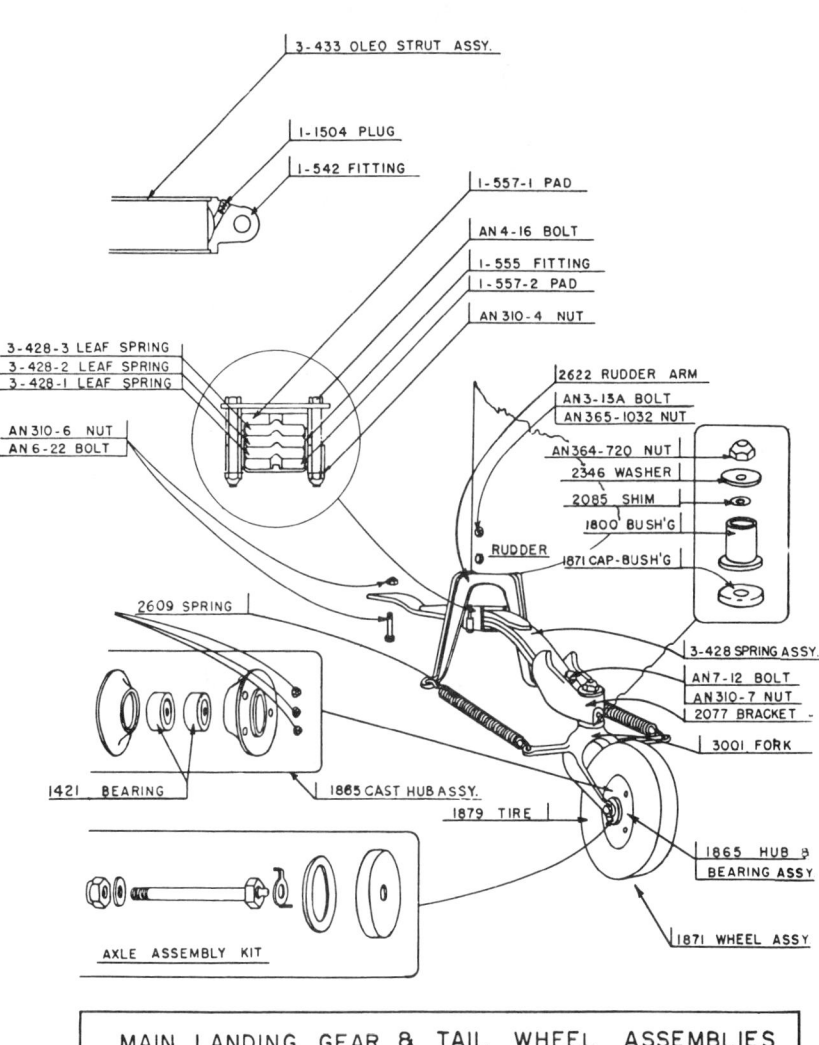

MAIN LANDING GEAR & TAIL WHEEL ASSEMBLIES

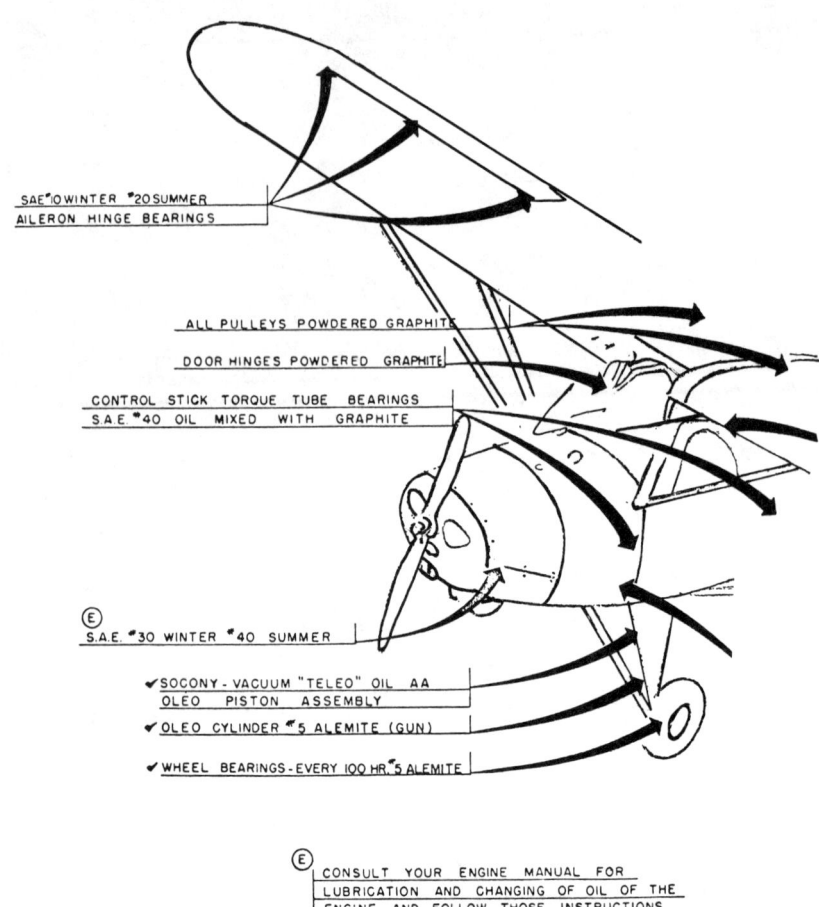

Fig. 6-16. A lubrication chart is a good thing to have after the project is completed, for regular ongoing maintenance of your airplane. All aircraft service manuals should have such a chart (courtesy Wag Aero).

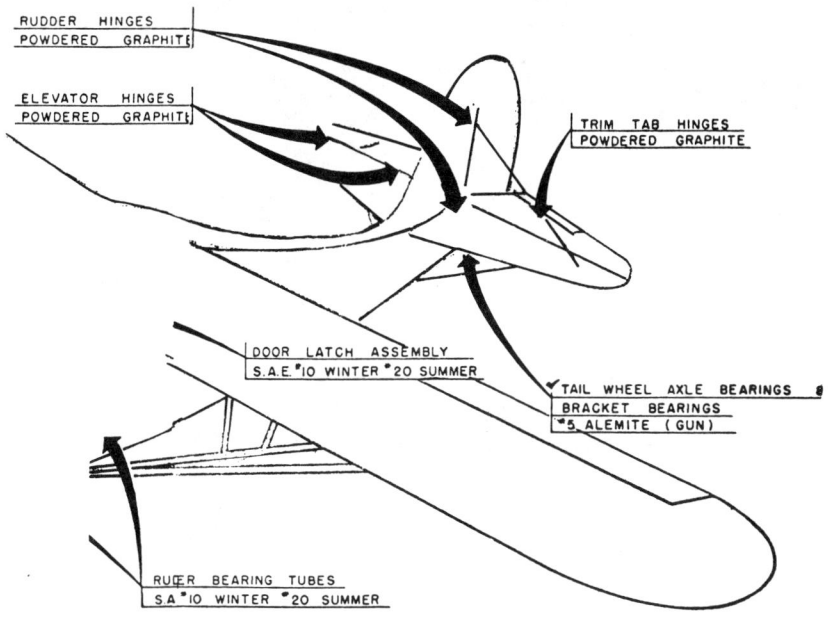

Fig. 6-17. Keep a notebook with a listing of the location of each cable that has to come through the fabric. You will need to know exactly where the hole should be cut in the new fabric. Measure in relation to a fixed object, such as the elevator mount.

the worst rust and get down to the underlying bare, shiny metal (Fig. 6-41). If you can do this at the worst places, you are virtually assured of having an airframe that will clean up fairly easily.

Just to be sure, however, there are other checks that you and your mechanic may wish to discuss. One way is to make a cut at a particularly bad rust spot and take out a short section of tubing. This

Fig. 6-18. Once the root fairings and wing root fabric is removed, the wing attachment bolts are accessible, along with fuel lines and electrical wires that must be disconnected.

Fig. 6-19. Tail group bracing wires are usually held in place by pins and safetied with cotter keys.

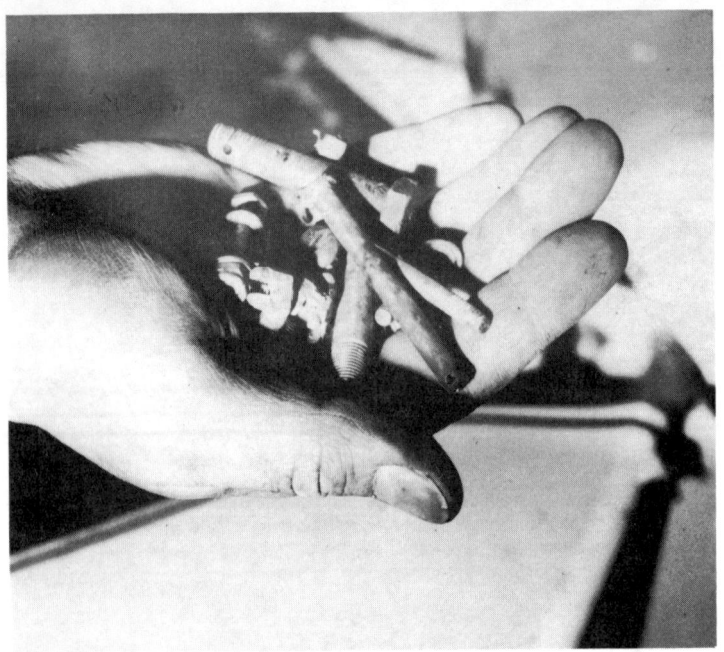

Fig. 6-20. Save all nuts, bolts and screws to make ordering new hardware easier. Replace all that are worn, stripped of threads or rusted.

piece, a mechanic (not you) will have to splice back in with a tubing "sleeve" and gas weld. Having made this cut, you will then be able to look inside the tubing to examine the extent to which rust has attacked the inside of the tubing.

Fig. 6-21. Once the wings are off, the airplane will fit through the shop door.

Fig. 6-22. Removing the side windows facilitates easier fabric removal, and also protects them from mishaps while working around the airplane.

Now here is one area in which having an older aircraft is very good. In the forties and fifties, the aircraft manufacturers filled the tubing of an airframe with linseed oil, allowed it to inundate the

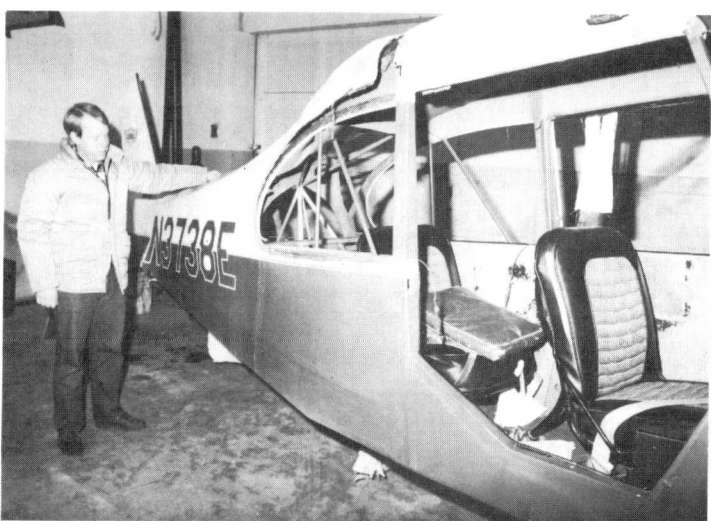

Fig. 6-23. Once the old fabric is cut down the middle of the fuselage, it can be easily peeled of, starting by loosening the fabric on top aft of the windshield. Fabric will then peel off with little effort.

85

Fig. 6-24. Off it comes! It is scary, in a way, to be beyond the point of no return on your recover project.

entire length of each tube, then poured it out and immediately sealed the ends with gas welds. Thus a film of linseed oil coated the inside of all tubes and prevented rust from getting started therein. Therefore, it is less likely that you will find rust inside the tubing of these older airframes.

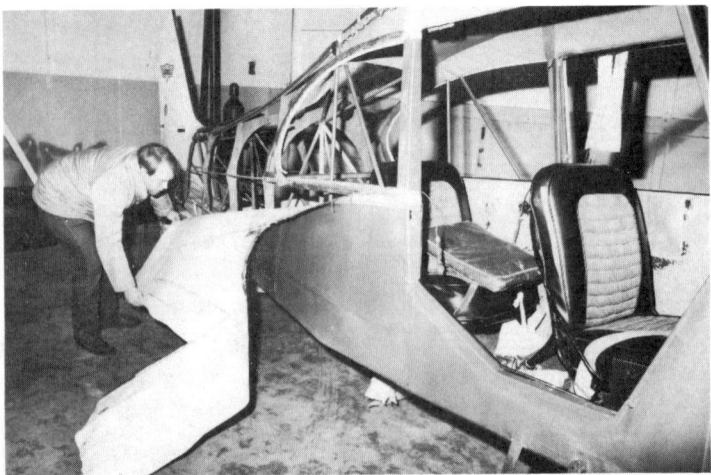

Fig. 6-25. As the fabric pulls away from the frame, carefully lay it down while additional cuts are made so as to avoid tearing the fabric sharply, possibly damaging salvagable wood beneath.

Fig. 6-26. As the fabric comes off, you'll get your first real look at the condition of the underlying structure.

If a sample has been cut from one of the worst rust blotches and no rust is found inside, the mechanic then can splice a new piece back into the cut and you can rest assured that your airframe is rust free. If

Fig. 6-27. This is an example of the type of screw fastener used to attach fabric down to underlying structural members. These are used in many light airplanes, but stitching and clips are used in others.

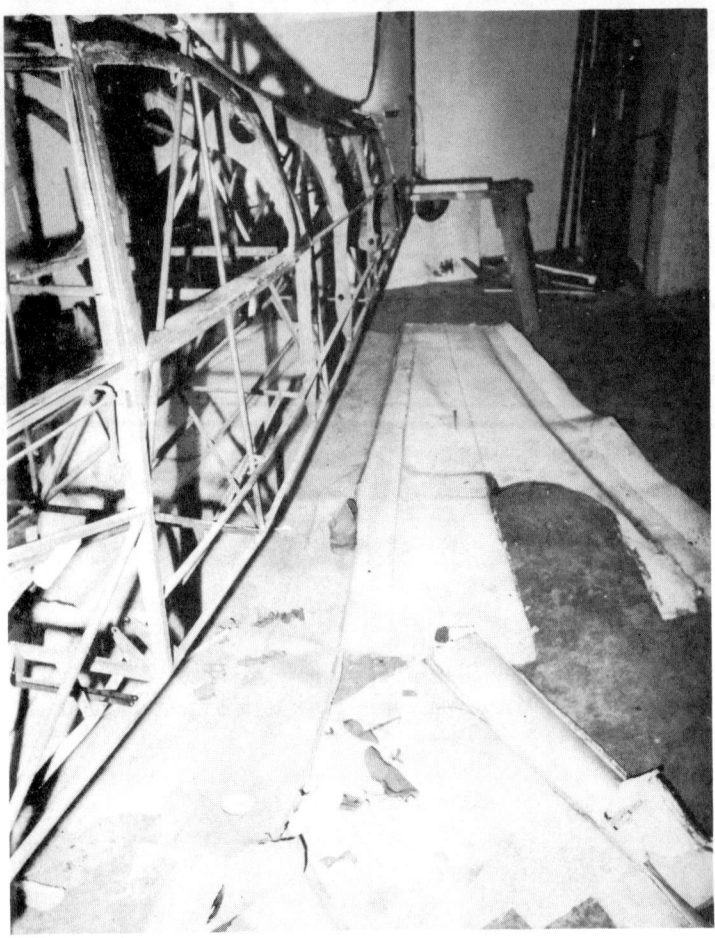

Fig. 6-28. Once the messy job of stripping the old fabric is done, roll up the old fabric and store it away for future reference as to the location of inspection plates, cable openings, drain grommets, etc. Then completely clean up the work area before proceeding.

rust is found inside, then it must be eradicated by splicing in new 4130 chrome moly (or whatever type was used in your particular aircraft) steel tubing wherever rust exists. Remember, rust is like a cancer in an airframe and will not repair itself. It will simply continue to eat away at your precious airplane and eventually destroy the metal entirely. In some extreme cases, the entire fuselage may have to be scrapped and an airworthy used fuselage (or new one, if one is available) be substituted. This is an expensive and difficult proposition, and one that should be assiduously avoided by the careful selection of a sound aircraft to begin with.

Assuming that you have found some rust and have determined that it is confined to the surface of the tubing, you are now ready to

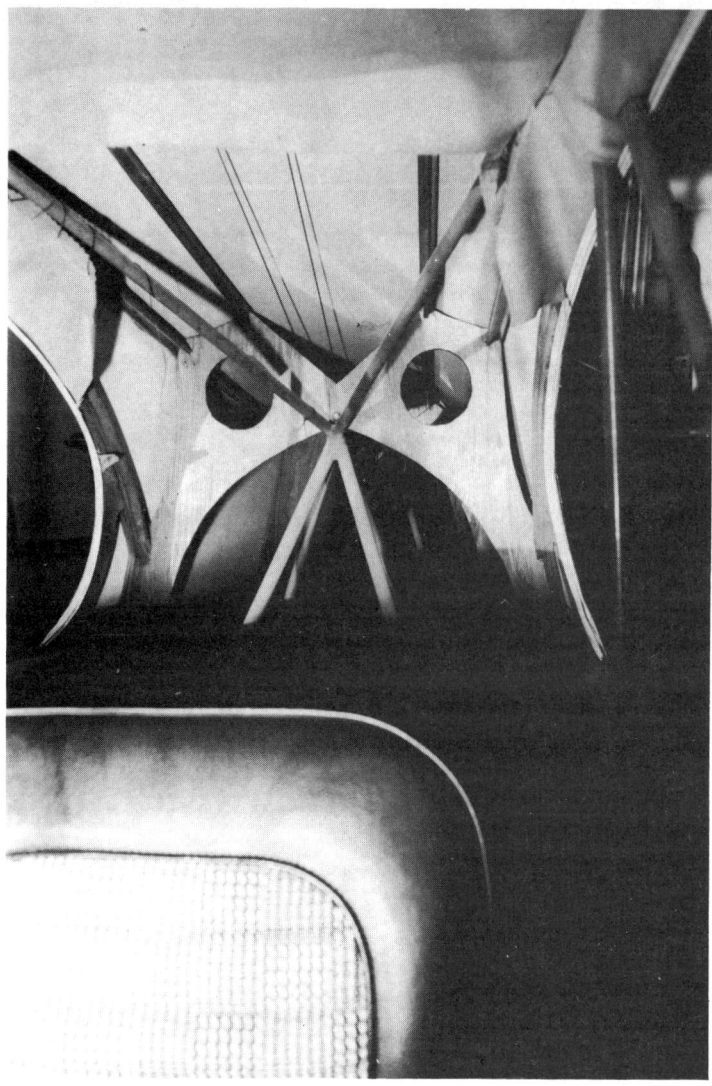

Fig. 6-29. Removal of the headliner is best accomplished by starting at the rear of the cabin and slowly working forward, unlacing it from the superstructure as you go. You will probably be replacing the entire headliner with a new one, since they can rarely be reused.

Fig. 6-30. Keep the work area as neat as possible by doing one thing at a time, thus avoiding the kind of mess you see here.

either thoroughly wire brush and hand sand the airframe or have it sandblasted. The former is done all the time and, provided enough long hours and elbow grease are applied, will get the job done. The latter is the best way to assure yourself (and your mechanic and the FAA) that you have thoroughly removed all rust and pitted rust or corrosion. Tiny particles of rust and corrosion left in the metal will be right back at work on the metal the instant you stop sanding.

To prepare the airframe for sandblasting, make sure that all wood is removed, along with every other item that can be taken off

Fig. 6-31. Carefully remove the windshield and store it where it where it will not be damaged. Before reinstalling it, polish it with "Mirror Glaze," or an equivalent Plexiglas polishing compound for a like-new appearance.

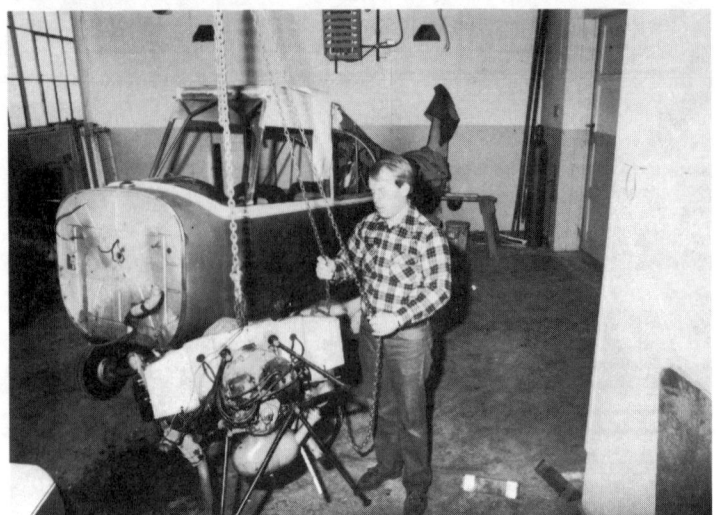

Fig. 6-32. An overhead hoist is a practical necessity for removing the engine.

the airframe. Tape over all threaded attachment shafts at the vertical stabilizer. This will protect these components as well as alerting the sandblaster to be careful with these brackets, fittings, etc.

Also be aware that only the steel tubing should be sanded. Never sand any aluminum components, such as wing ribs or formers.

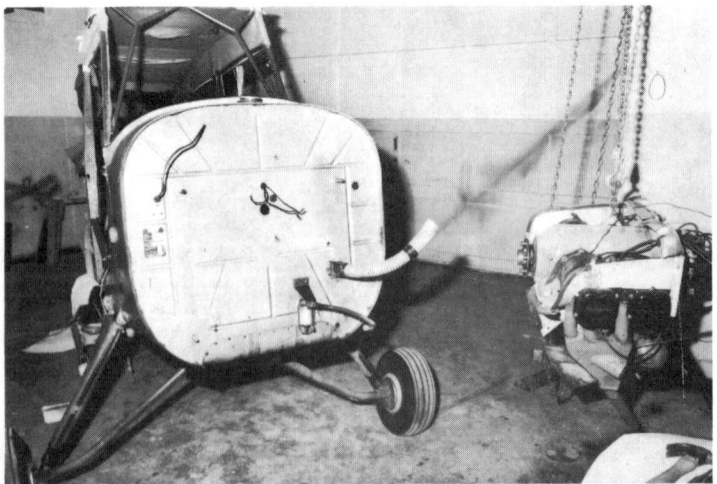

Fig. 6-33. Once the engine is off, the rest of the firewall fittings and the firewall itself will easily come off.

Fig. 6-34. A nose-mounted fuselage fuel tank, such as the one on the Aeronca Champ, can be easily removed by unbolting the metal band that holds it in place.

Having your airframe sandblasted will probably involve having to take the airframe to a shop somewhere for the work. Some sandblasting firms have portable units that can come to you, but if

Fig. 6-35. This wood has rotted through completely from long years of exposure to moisture. When this much rot is found, the entire wood structure comes under suspicion and warrants a very thorough inspection by your mechanic, or replacement with a new wood kit.

Fig. 6-36. When the fabric comes off, you will be confronted with the condition of the metal and wood. Don't let this first look scare you; this metal and wood is probably salvagable.

this is not the case, be prepared to have a small trailer, or at least a long-bed pickup truck to haul the airframe in. Adequately cushion it against bumps or jolts during transport so that you; will not crack or

Fig. 6-37. Time and the elements will probably have buckled the zinc chromate primer, allowing some degree of rust to attack the metal tubing. In this case, sandblasting is preferable, but a thorough hand-sanding job might suffice.

bend airframe members. Also make your arrangements with the sandblasting firm to have all the work done in one day. It will usually take three or four hours. Then, be there to pick it up when they are finished. I once heard of a person who took his airframe to the

Fig. 6-38. Remove the old wood structural fairings by unbolting them from the metal fuselage frame. If a diagram is not available for your airplane, make a careful sketch of the wood structure, method of attachment, etc.

Fig. 6-39. Rust may be evident on the metal tubing. A stiff wire brush can be used to knock off much of the old primer, rust, and dirt, in preparation for sandblasting or hand-sanding.

sandblaster and left it for a few days while they worked on the wings and control surfaces. Upon returning to the sandblasting shop, he found that his airframe had been finished on the first day, and then placed outdoors in the back lot for the rest of the time. A bare metal,

Fig. 6-40. Down to the bare airframe, the airplane is now ready for sandblasting.

freshly sandblasted airframe has *no* business sitting outside in the elements. To allow this is asking for more rust and corrosion to set in on the unprotected metal.

You must protect the airframe from the elements by immediately taking it to your dry workshop. You will therefore want to immediately take your airframe to a dry place and make very sure that all metal surfaces are free of dust, dirt, and moisture.

Once the sandblasting is completed and the airframe is back in the workshop, carefully, with your mechanic, check the whole thing over thoroughly for cracks in the metal. It is extremely important that this inspection be conducted prior to priming the fuselage, as it will then be too late. You may even want to use a spray such as "Dye-Check" to make sure that no tiny hairline cracks exist. These crack-testing products normaly use an aerosol cleaner, a dye and some type of aerosol chemical developer. The metal is cleaned,

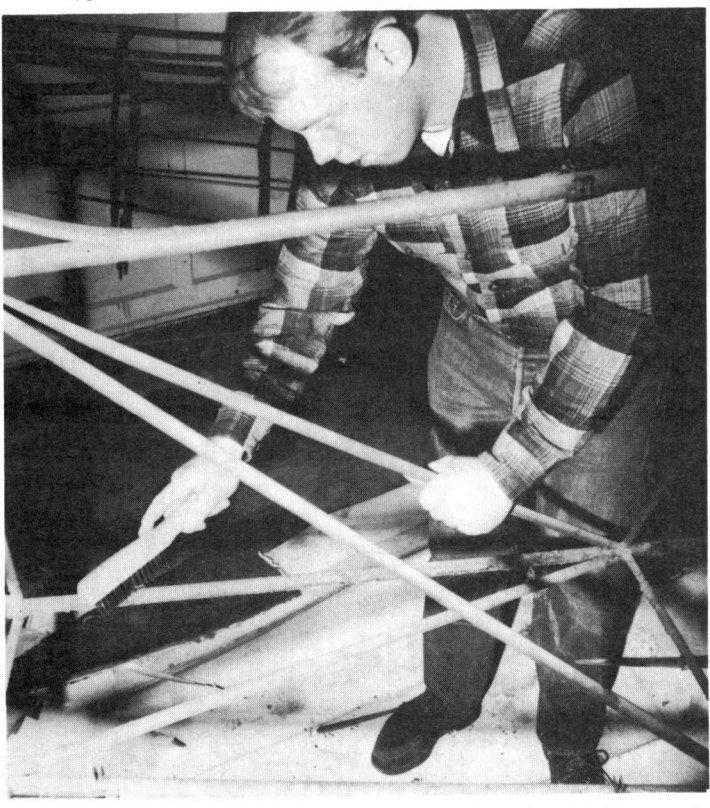

Fig. 6-41. Go over the entire tube structure with a stiff wire brush, taking care to make brush strokes parallel to the longitudinal axis of the airplane.

then the dye is sprayed on and allowed to sit for a few minutes. Then the dye is wiped off and the developer is sprayed onto the metal. The developer is usually light in color, and will pull the dark-colored dye back up out of any cracks in the metal. The crack will then show up as a dark dye line against the lighter surface. Though dye-checking is not required, it is a fairly inexpensive way to make sure the metal is completely free of cracks or breaks. Be sure to carefully reclean the metal with acetone, or other non-petroleum based cleaner prior to painting with anti-rust/corrosion spray primer.

Zinc chromate primers may be used as they have adequate rust/corrosion inhibiting qualities. They are all right to use, but better substances are on the market now. Some are epoxy-enamel primers, others are tough zinc-based primers that are supposedly impervious to rust and corrosion. Check with your mechanic for the type he recommends.

Go over every inch of metal with at least two coats of primer (Fig. 6-42). I might also mention here that whatever primer you use as a rust and corrosion inhibitor should also be dope proof. In other words, the paint you use has to be able to withstand the dope with which you will be attaching the fabric to the tubing. If the dope will buckle and remove the rust-inhibiting primer, it will not be there doing the job, and you will not know it since the metal will be covered up at that point. So be careful to select an aircraft metal primer that is compatible with the dope and fabric nature of your airplane.

Once the sanding and priming of the metal frame is complete, you have reached a turning point in the project, at least as far as the fuselage is concerned—you are ready to start putting things back together again.

Let us set aside the fuselage for now, and take a look at what will be done to disassemble the wings and control surfaces and prepare them for recovering.

Preparing Wings and Control Surfaces

If you can build a rotating stand that will accommodate a wing attached at the root and supported on the tip by a regular sawhorse, you will be able to turn the wing with ease as you work on it. If such a stand cannot be built or obtained, simply support the wing with standard sawhorses (Fig. 6-43). Being able to turn a wing quickly over is a definite advantage and a big time and work saver.

Once the wing is established in the work position, begin removing the old fabric.

Fig. 6-42. Once you have sandblasted the airframe, immediately apply several coats of zinc chromate or other rust-preventive primer to prevent the onslaught of every airframe's enemy: rust.

The old fabric is usually harder to remove from wings and control surfaces than from the fuselage, because of the method of attaching the fabric to those surfaces. Your airplane probably employs either rib stitching, screws with washers, or clips to fasten the fabric to underlying ribs or structural members. Regardless of the method, to properly and carefully remove fabric from wings and controls is a painstaking and time-consuming task.

The best method I have found is to begin by peeling off the tapes that cover the fasteners. Once these tapes are loosened at either end, they tend to peel right off, exposing the underlying fasteners. The older the fabric, the easier they usually come off. Once these overlying tapes are removed, the task of cutting the rib stitching or removing the screws or clips can begin.

In all likelihood, unless the airplane is really far gone, these fasteners (clips or screws) will be reusable. *Keep all unrusted rib fasteners*. Believe it or not, these little items are expensive in the numbers required to assemble the airplane. There are three or four hundred, maybe more, on each wing. Carefully place them in a well marked container for later use. Throw away broken, rusted or corroded fasteners and keep track of how many you'll need to reorder at around $.05 each. After you have removed all fasteners, the fabric will peel off very easily, leaving you with the basic wing structure, a lot of old fabric, dust, a few birds nests, assorted beehives, and the faded and forgotten domiciles of the ubiquitous mud dauber. *Save the old fabric* so you'll know where to put the inspection plates for the controls and other areas where inspections are required. See your aircraft service manual for the locations of the

Fig. 6-43. Place the wing on sawhorses at a convenient working height. A rotating stand is better, if one is available or can be made.

Fig. 6-44. The wing ribs, along with the rest of the wing structure, most conform precisely to their designed shape, or the airfoil will not perform as intended.

inspection covers, too, as most of these manuals list such information.

When the fabric is all stripped away, blow the dust out of the wing structure with the air hose and then, once everything is fairly clean, take a good look at every component of the wing.

Your mechanic will want to closely examine the spars, ribs, drag wires, compression members, control routes and pulleys, and, in general, the overall condition of the interior of the wing (Fig. 6-44).

If your wings are in such good condition that no rust is present on the compression members or drag wires, and the spars are nicely coated with varnish and free of cracks, and the wood areas around all metal attachments to the spar are clean and free of dark spots, you might be able to clean the wing structure and have it ready for new fabric. However, it is a distinct possibility that some rust will be evident on the metal members, or that the spar attachments warrant a more thorough check. If this is the case, then you will want to remove the metal compression and drag wire members, the ribs, woven alignment tapes, and completely examine each spar. When you do this, you have disassembled the wing, and a mechanic will have to aid you in preparing the wing for reassembly, reassembling and trammelling the wing and reattaching all members and fittings. It is an exacting job, and one I will not detail here. It requires that you measure precisely the relationship of the ribs to one another and to the spar; you will have to make sure that you reinstall all bolt attachments without overtightening them and damaging the wood spar. The drag wires must each be measured for trueness and the reinforcing tape must be rewrapped. In all, it is a task that a mechanic should be involved with from start to finish. I would recommend that you read *Aircraft Dope & Fabric*, by Ruth and Warren Spencer, (TAB Book No. 2313 Blue Ridge Summit, PA) for a good description of the process involved. Also consult AC 43.13 for this and any other procedure involving the dismantling and reassembly of airframe components.

When each wing is properly assembled and trammelled, with the wooden spars carefully sanded and adequately varnished with spar varnish, your mechanic will be able to make his final inspection and sign off the wings for cover.

At this point, both wings, and the control surfaces, most of which will be constructed of 4130 and smaller tubing and treated much like the fuselage metal frame, are ready for fabric, along with the fuselage. You will now begin "dressing" the structures in the fabric envelopes.

Fig. 6-45. Tip: If you own an Aeronca 7AC or other aircraft that requires a "dorsal fin" conversion for the addition of a higher horsepower engine, now would be a good time to do the conversion, since the fabric is off. It will eliminate any obstacle to adding higher horsepower later, plus increase the resale value of your airplane.

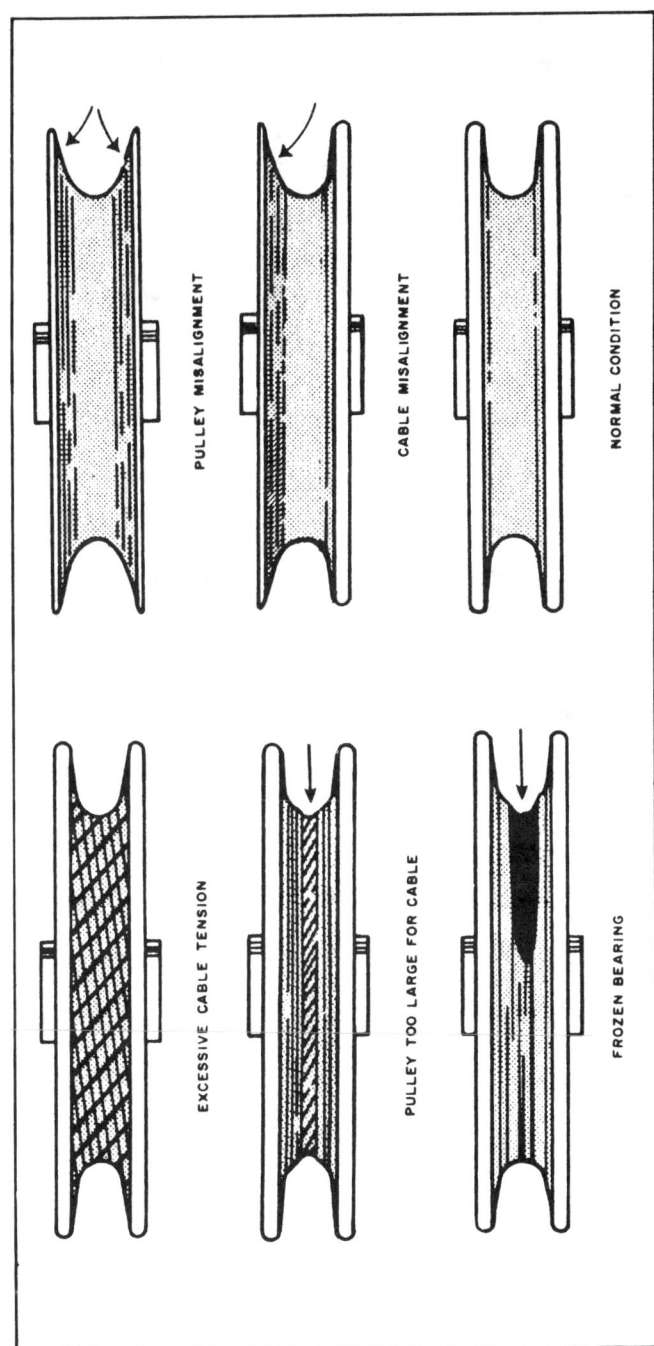

Fig. 6-46. This diagram, taken from FAA Advisory Circular 43.13, shows pulley wear patterns and their cause. A careful inspection of all control pulleys is required prior to reassembly.

Preparation for Fuselage Cover

Now that your fuselage is completely painted and ready for reassembly, it is at this point that you will replace the wood formers, if any, that make up the shape and structure of your airplane. As I have mentioned, complete wood kits are available from such companies as Wag Aero, and aside from a complete kit, replacement wood parts are available from, among others Aircraft Spruce & Specialty Company (Box 424, Fullerton, CA 92632, telephone: (714) 870-7551). You can either replace or repair the wood and fasten it on the airframe exactly as it came off. Have your mechanic supervise the installation of the wood.

While installing the wood members, some pointers I can give you include making sure that you carefully tighten attachment screws and bolts. Never overtighten a screw in wood, as it will crack or crush the wood and render the attachment unsound. Also, be sure that once the wood is all attached and tightened in accordance with the aircraft service manual and your mechanic's instructions, that it is sanded very smooth at the points where fabric will touch it, and that a good coat or two of spar varnish is applied to the wood and allowed to dry. Once the wood has been thus treated, it is a good idea to apply masking tape to the top surfaces of the wood where the fabric will rest.

In all of this, consult the FAA and your mechanic supervisor along with the requirements and procedures of AC 43.13 to determine exactly what must be done and who must do it. Extensive woodworking and repair are beyond the scope of this book and it is doubtful that much will be required, providing that you choose carefully from among the more standard, popular light aircraft (Fig. 6-45).

Once the wood formers are in place, and all control channels and pulleys replaced, the fuselage is basically ready for cover when all control cables and wires have been put in place with their appropriate brackets and control cable guides. Electrical wires should be routed aft through "Adel" type clamps that have a rubber liner inside a metal clamp. These clamps eliminate friction and fraying of wires.

Make sure that all elevator, rudder and trim controls are properly in place and routed through the appropriate channels and guides. Remember that once the fabric envelope is placed over the fuselage, all this will not only vanish from view, but will be completely inaccessible.

Lubricate all pulleys and controls and check them for free and easy operation. Replace bushings that are worn, as well as any

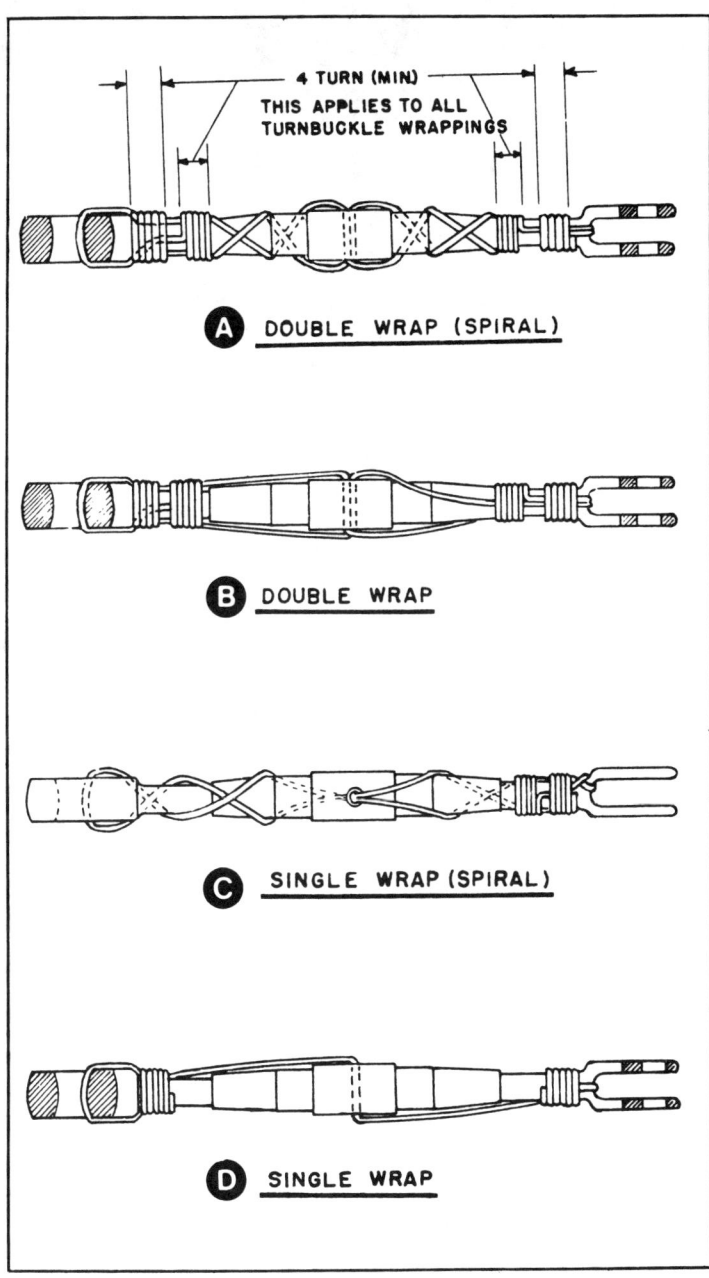

Fig. 6-48. Turnbuckles are used to connect aileron and other control cables at various points. This diagram gives methods of safetying turnbuckles, from AC 43.13.

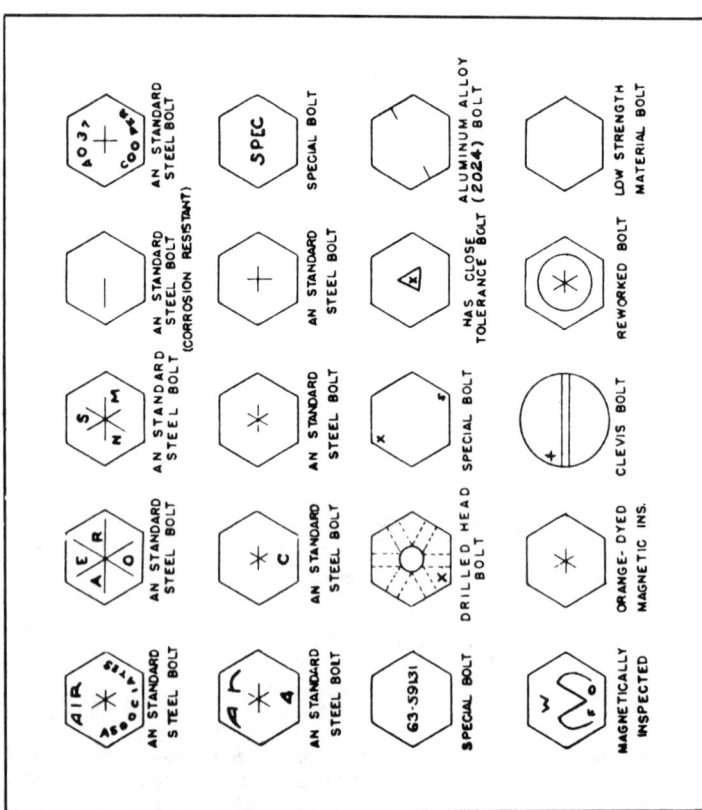

Fig. 6-49. This chart, from FAA Advisory Circular 43.13, will help you to identify the type of hardware required in the airplane. It is absolutely essential that you replace aircraft bolts with approved "AN" aircraft hardware!

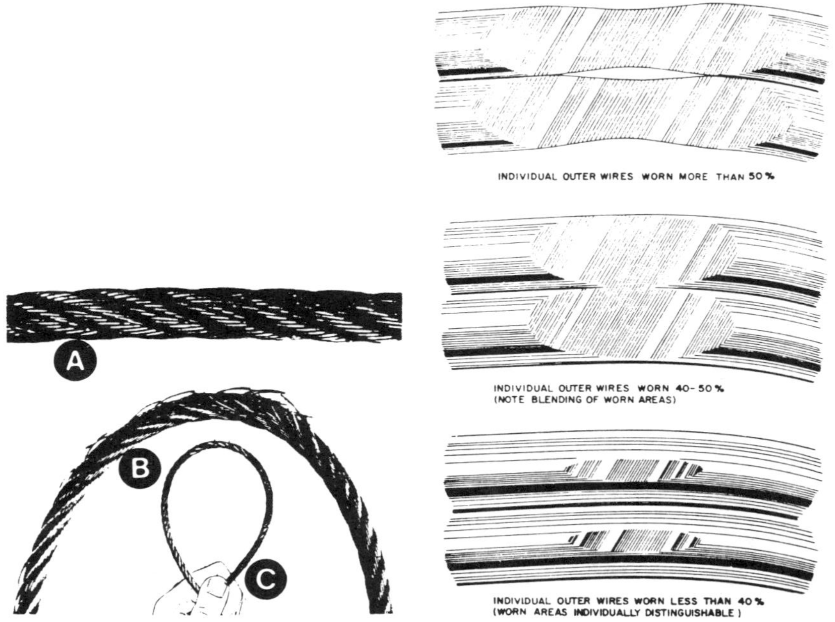

Fig. 6-50. Before reinstalling any control cable, make sure that it has been inspected and conforms to the criteria of AC 43.13.

control pulleys that are worn out or unevenly worn. Uneven wear on pulleys usually indicates poor rigging of the cables or the pulleys themselves, or perhaps worn out bushings or bearings in the pulleys. Control cables should track straight and operate smoothly with no binding or dragging. This is not only a safety consideration, but will have a great bearing on the feel of the controls and the handling of the airplane. Great care, along with the supervision of a mechanic is advised where hooking up, checking and safetying of control cables is concerned (Figs. 6-46 through 6-50).

Once all control cables, wires, etc. are in place according to the aircraft service/overhaul manual, tape all cables and wires up well inside the fuselage frame with masking tape. They will be retrieved later through inspection holes and control cable openings and attached in place, but for now, you want them fixed out of the way of the fabric work which is about to begin.

Your fuselage should now be ready for its precover inspection. The FAA, or your mechanic/airframe inspector must then sign the fuselage off as "ready for cover," which means that you may procede with the application of the fabric.

Chapter 7
Applying the Fabric

As an amateur rebuilder, I recommend the use of pre-sewn fabric envelopes for use on your light aircraft. They are much easier to work with because the complicated measuring, marking, cutting, and most of the sewing has already been done. Please note that word *most*. No matter how good the fabric envelope is, it seems there is a little sewing to do. For one thing, the ends have to be sewed or doped closed, but frequently, a pre-sewn envelope will have to be cut in order to get it over the vertical fin, for example. This varies from airplane to airplane, but the sewing you will have to do with an envelope is much less than that you would have to do using the so-called "blanket" method.

To clothe the airplane in fabric envelopes, keep in mind that the fabric must always be placed on the airframe so that it conforms, to the greatest possible extent, to the way in which the original fabric was applied (Figs. 7-1, 7-2). Use the instructions provided with the fabric process you choose to use. Also, always consult the FAA Advisory Circular AC 43.13, which you should have a copy of prior to starting your project.

The procedure for installing Ceconite in the approved manner is given here. Ceconite Procedure Manual 101, along with its appendices, is reprinted here, courtesy of Ceconite, Inc.

This is a good time to re-emphasize the fact that in order for a recovery job to be legal, it must be FAA approved. This means that whatever process for recovering you choose, you must make sure that it is FAA approved and that the manufacturer or supplier of the fabric, dopes, and other materials and processes holds a Standard

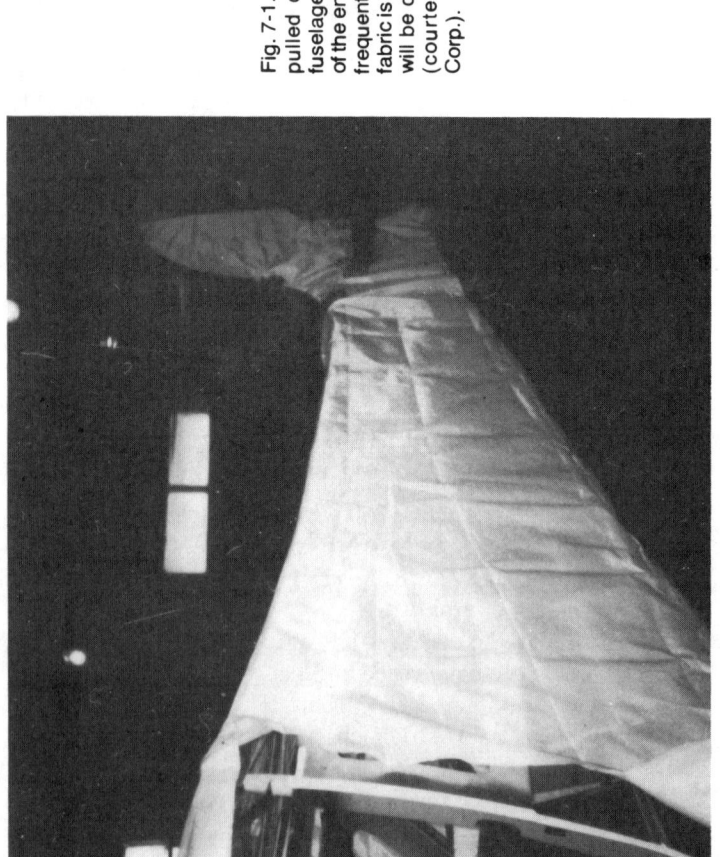

Fig. 7-1. A Ceconite envelope is pulled over this Taylorcraft F-19 fuselage. Getting the narrowest part of the envelope over the vertical fin is frequently the hardest part. Note that fabric is loose at cabin door frame. It will be doped securely to the frame (courtesy Taylorcraft Aviation Corp.).

Type Certificate, commonly referred to as an STC. All of the fabric covering methods discussed here in this book are FAA approved—Stits Polyfiber, Ceconite, Copper Dac-Proof System, Razorback, Airtex, etc. Beware of any advertisements offering low prices on bulk materials, such as fabric and dope. These are not very likely to be FAA approved and are also likely to lack the strength and quality requirements of FAA approved materials and substances. One quick way to make sure that your fabric is FAA approved is to look for a dark thread running through the fabric at three-foot intervals. This is the mark of fabric that was manufactured under FAA approval. Some fabrics will bear a mark such as "FAA-PMA Approved," or words to that effect, identifying them as approved fabrics.

Follow the instructions for your procedure exactly. The Ceconite Procedure Manual 101 is given here only as an example of the method of covering an aircraft with Ceconite fabrics. Though similar, the methods for covering Grade A cotton, Irish linen, and others are different. In any case, your mechanic supervisor should be available to consult with you, answer any questions as the work proceeds, and, of course, sign off your work upon completion.

Ceconite Process—General

The aircraft and finish procedure described herein is devised from results listed in Ceconite, Inc. Report No. 101. The procedures of covering and finishing are simple and may be accomplished by any A&P mechanic or individual having previous aircraft fabric experience. In the absence of specific instructions, the techniques, procedures, and directions of FAA Advisory Circular 43.13—1A and 2 will govern.

Since the days of the JN4D (Jenny), the fabric-covering process of aircraft has undergone virtually little or no change. As a result of years of experimentation, Ceconite, Inc. has developed a synthetic fabric (Ceconite 101) which is a major advancement in fabric coverings. For the same weight, Ceconite 101 is much stronger and far more durable than cotton TSO C-15. Thus, for the first time it is possible to enjoy the advantages of a lightweight fabric covering together with probable lifetime durability.

Ceconite 102 is a fine weave medium weight (2.8 oz. per sq. yd.) fabric, having tensile properties slightly greater than Grade A cotton. It is considered a standard replacement for Grade A when smooth finish or weight saving is of consideration. Ceconite 103 is a fine weave lightweight (1.7 oz. per sq. yd.) fabric for use as a finish

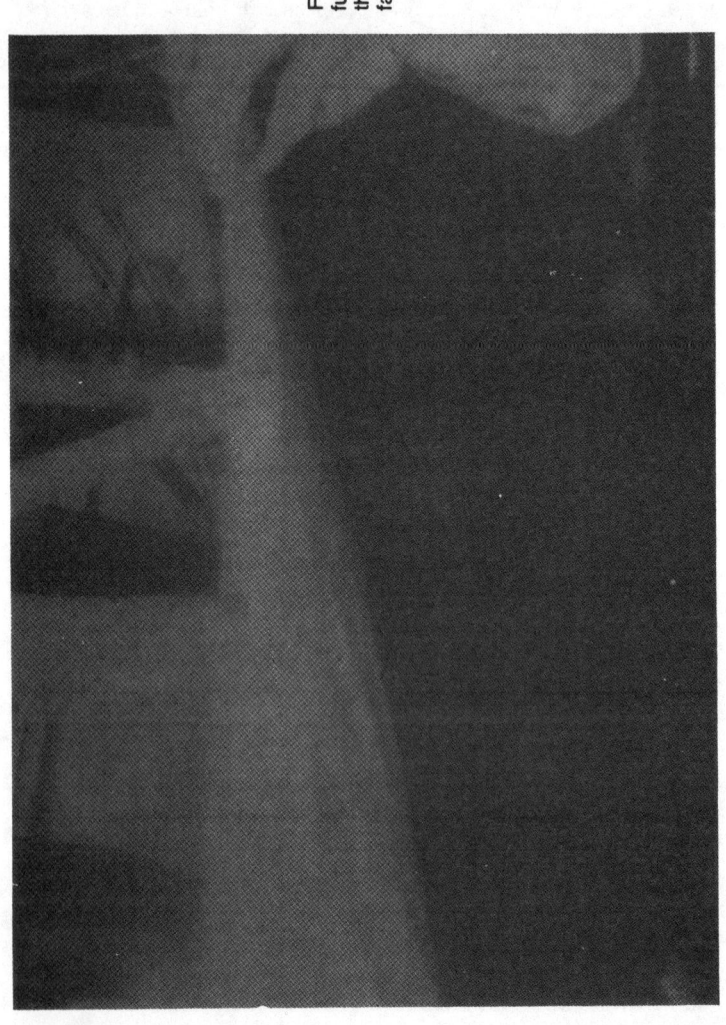

Fig. 7-2. Covering his Aeronca fuselage, this owner is employing the "blanket method," cementing fabric to the fuselage in four pieces.

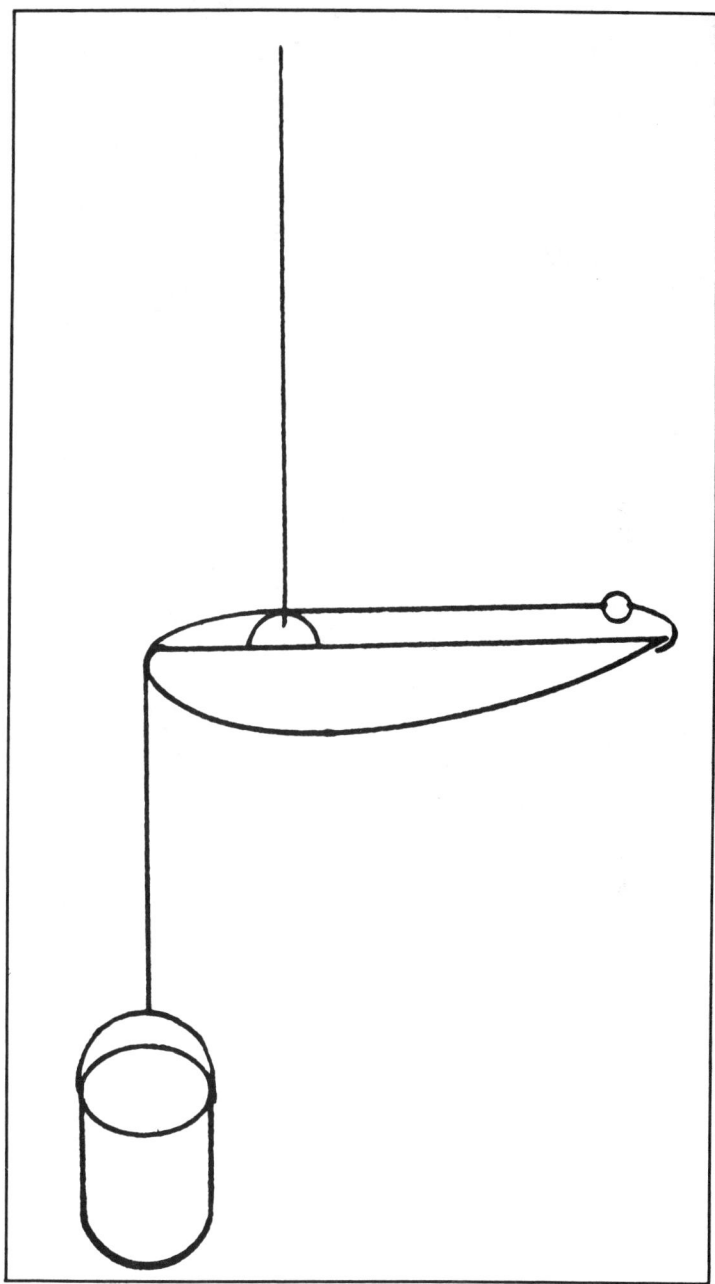

Fig. 7-3. This simple method for checking the center of gravity of control surfaces is reprinted from the Ceconite Procedure Manual #1 (courtesy Ceconite, Inc.).

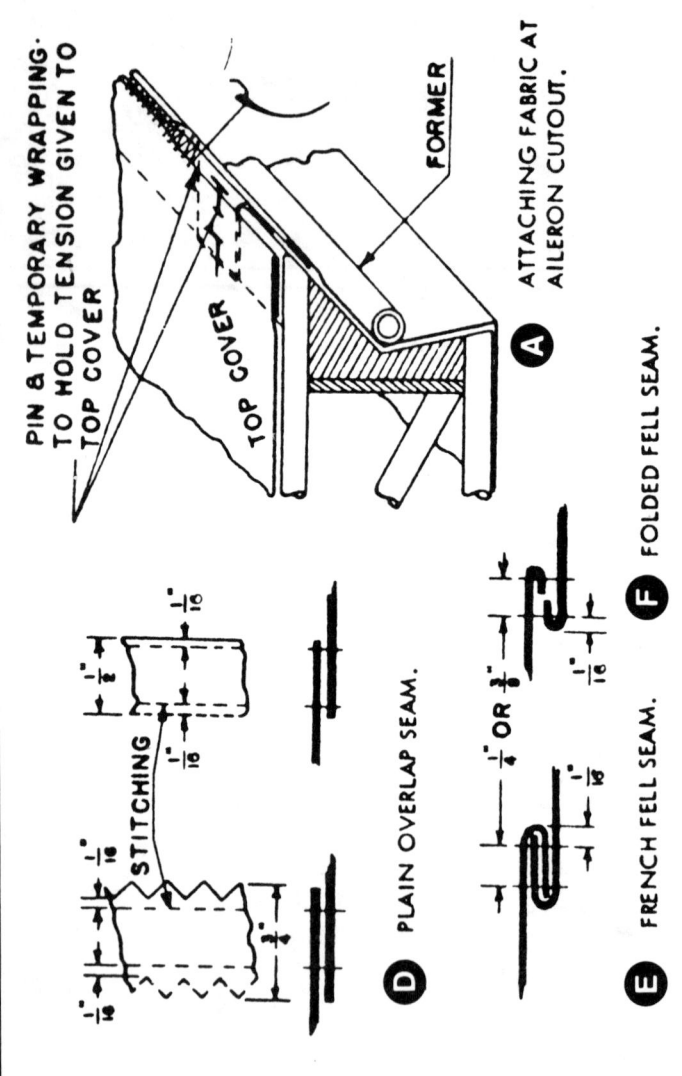

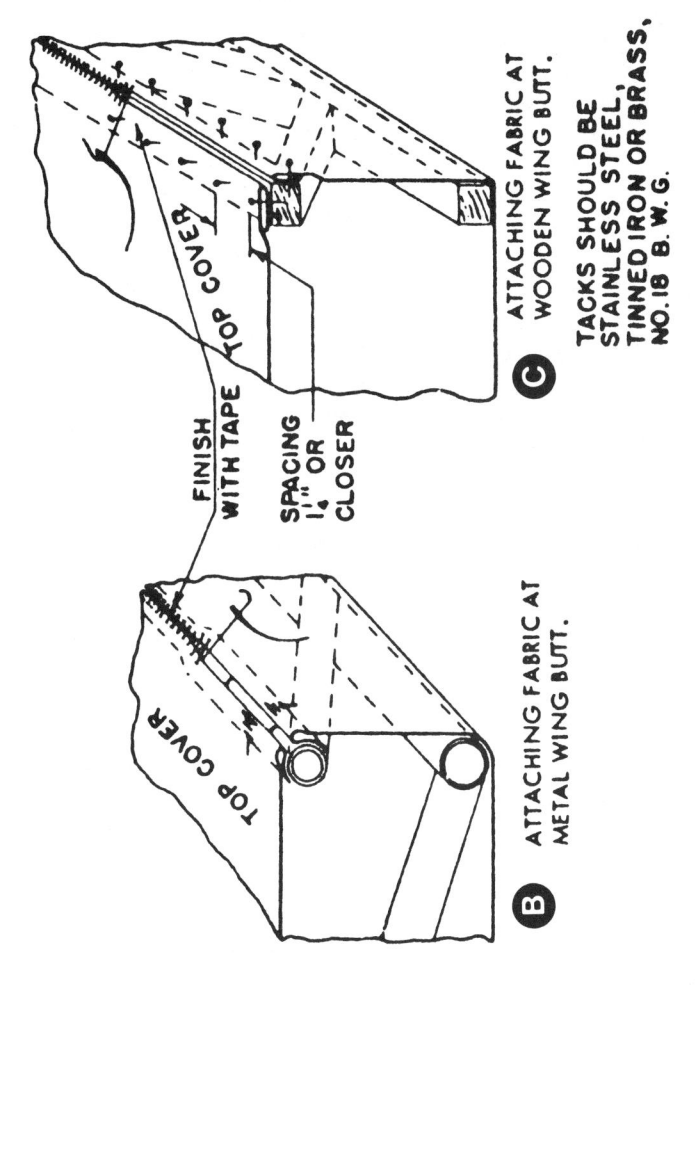

Fig. 7-4. Typical methods of attaching fabric.

over areas covered with plywood. Ceconite 103 may not be used as a primary structural material except for sailplanes and gliders. FAA approved Ceconite has Ceconite 101, Ceconite 102 or Ceconite 103 stamped on the selvage at one yard intervals. Ceconite process STC numbers are specifically restricted to the users of Ceconite and may not be used for other processes. These STC numbers may be used free of charge by purchasers of Ceconite fabric. In addition to the FAA approval stamp, Ceconite 101 has three longitudinal marker yarns woven in for positive identification.

1. In cases where other than a standard finish on control surfaces could possibly be altered, the control surfaces must be rebalanced in accordance with the aircraft manufacturer's instructions. See Fig. 7-3 for method of checking control surface CG.

2. Since Ceconite 101 and our standard finishes weigh identically the same as aircraft cotton and usual dope finishes, no change in CG or weight will occur.

3. When old cover is removed: (a) Write down (and sketch) method of securing fabric to parts. (b) Save old cover for use as a pattern for locating openings, grommets, inspection rings, determining rib cord lacing areas and spacing, etc.

4. Prior to attaching fabric: (a) Make necessary repairs to metal and wood members. (b) Replace electrical wiring that will not last at least 10 years or will be inaccessible after the aircraft is recovered. (c) Replace any cables, pulleys, guides, controls, brackets, formers or items showing signs of wear or deterioration and not capable of lasting at least 10 years. (d) Zinc chromate metal surfaces or paint with Rustoleum. (e) Refinish and reseal wood surfaces as required. (f) Smooth or tape all rough or sharp edges to prevent chafing through fabric covering (Fig. 7-4).

5. Ceconite covering methods: (a) Envelope and blanket method: envelopes or blankets must be installed in the identical manner of the manufacturer's original envelope with particular attention to attach method, rib cord lacing areas and spacing (or screws, or clips, depending on the particular aircraft). Machine sewn seams, baseball hand stitching, lock knots, sein knots, seam directions, and overlap areas must duplicate the manufacturer's original covering or adhere to FAA Advisory Circular 43.13-1A and 2 when deviating from the original covering. Either "Super Seam" or an aviation cement will be used in areas calling for cemented joints. Dope may be used for laying surface tapes, etc.

On some low-speed aircraft which do not exceed the "never-exceed speed" listed in FAA Advisory Circular 43.13-1A and 2,

Ceconite 101 may be attached to the longerons using an FAA approved cement.

(b) Cemented method using "Super Seam Cement:" The cemented method using "Super Seam Cement" is not restricted to aircraft which never-exceed speeds of 150 mph or under, and therefore, may be used on any fabric-covered aircraft or fabric-covered control surface. The cemented method of covering is recommended on all aircraft where fabric width is sufficient to cover an entire side of a surface. If this is not possible, the fabric may be run spanwise on one side and cemented with "Super Seam Cement" to the leading and trailing edges as well as to the root ribs and wing tips. The opposite side is then applied using "Super Seam Cement" with at least a 4" overlap at the leading edge, a 3" overlap at the trailing edge, and at least a 1" overlap at the top. Surface tape of 6" width should be cemented over the leading edge overlap and at least 3" width tape cemented over the trailing edge and wing tip overlaps. On the fuselage, fabric should be applied on opposite sides with "Super Seam Cement" then applied using cement with at least a 1" overlap over the longerons and fabric previously applied to the adjacent sides. Surface tape at least 2" in width is then cemented over the overlap seams using "Super Seam Cement."

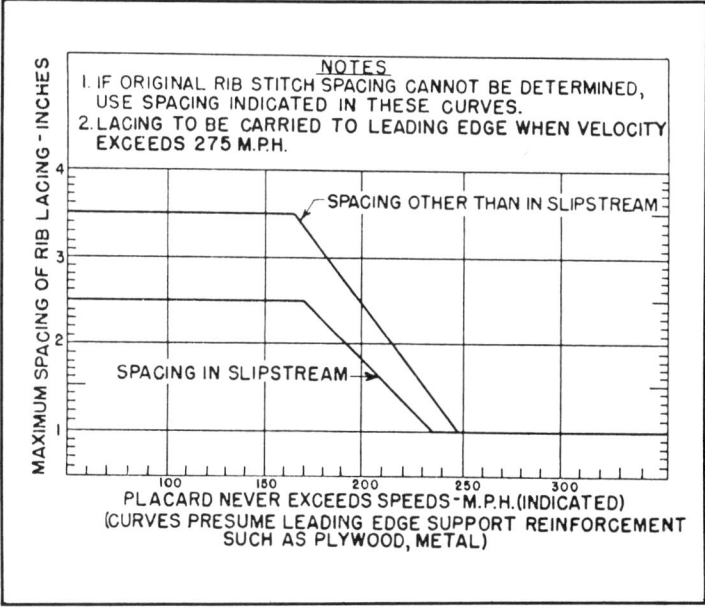

Fig. 7-5. Fabric attachment spacing.

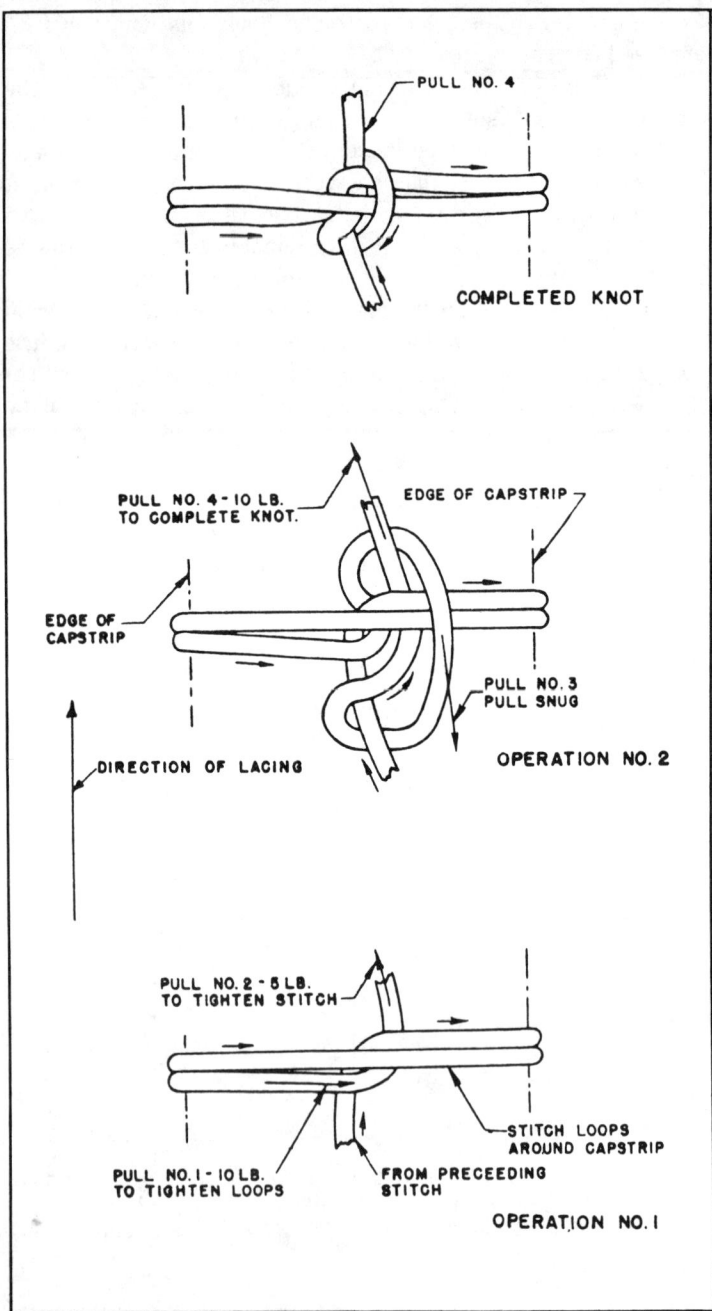

Fig. 7-6. Standard knot for double loop lacing.

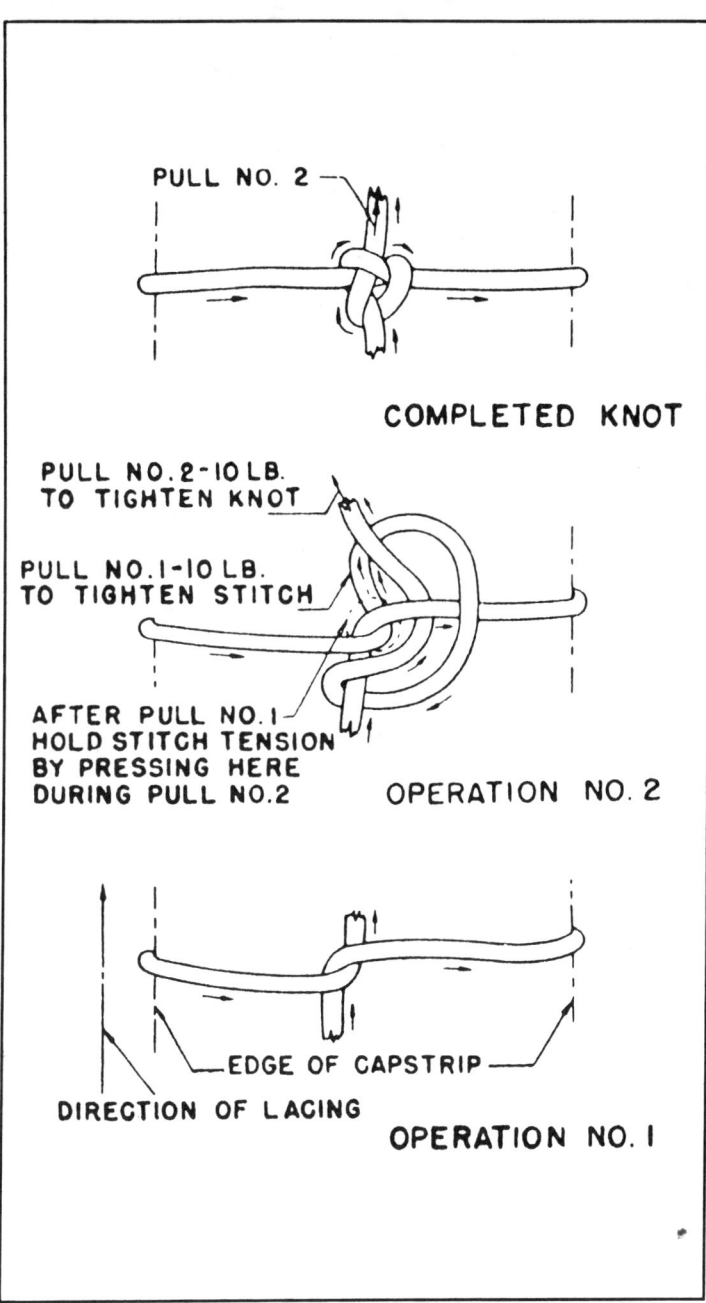

Fig. 7-7. Standard knot for rib lacing and terminating a sewed seam (modified seine knot).

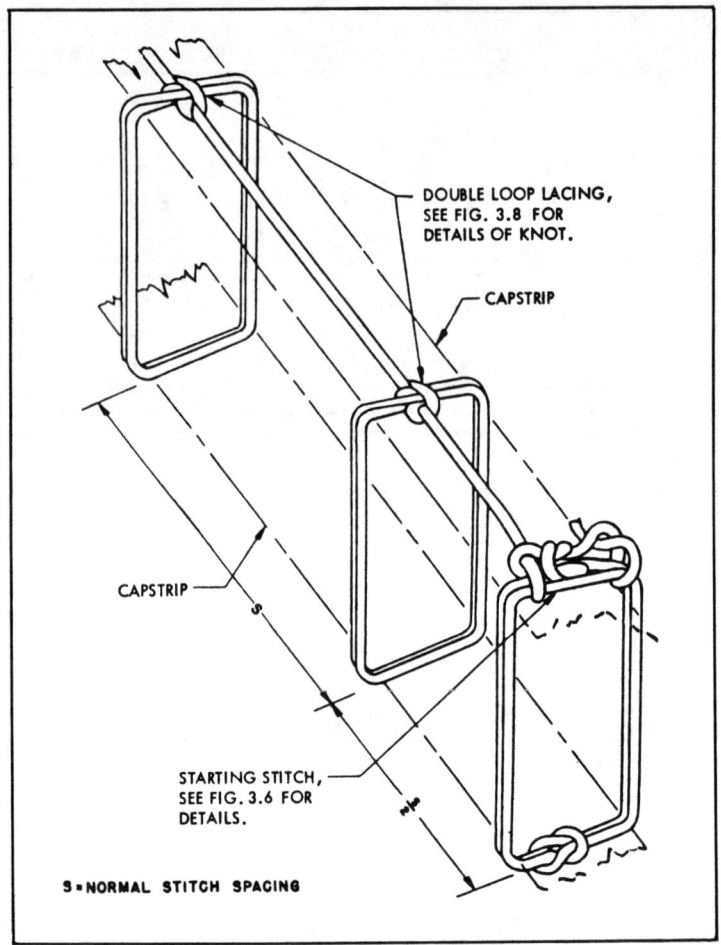

Fig. 7-8. Standard double loop lacing.

6. Rib lacing cord, reinforcing tape, etc.: "Ceconite Process" D-693 or D-415 Rib Lacing Cord, D-69 Machine Sewing Thread, and D-207 Hand Sewing Thread will be used for rib lacing, machine sewing and hand sewing. Surface tapes may be either predoped cotton (TSO C-15), Ceconite, or "Air Fibre." Reinforcing tapes may be either "Special Ceconite Process" or cotton. If cotton tapes are used it should be understood that a comparable fabric to tape life expectancy can not be expected.

7. Degree of tautness in attachment of Ceconite 101: In either method 5.a. or 5.6. preceding, Ceconite should be attached to the structure slightly looser than when covering with cotton fabric. Do

not become concerned about fabric wrinkles as they will vanish upon application of heat. Ceconite 101 is attached to the airframe and given an initial shrinking to first remove wrinkles and general slackness. Final shrinkage (see 8, below) is then accomplished following which rib lacing is accomplished. The covering is now ready for installation on surface tapes, inspection rings, reinforcements and grommets. The area under the surface tape must be given two heavy coats of dope before applying surface tape.

8. Ceconite is shrunk by heat. In general, Ceconite shrinkage is in direct proportion to the degree of applied heat, 400°F will shrink unrestricted Ceconite approximately 10% or 5" in 50" width. Coverings for gliders and light frame aircraft should initially provide approximately 1" slack per 50" to preclude structural warpage. Ceconite 101, when subjected to 240°F (household flat iron on

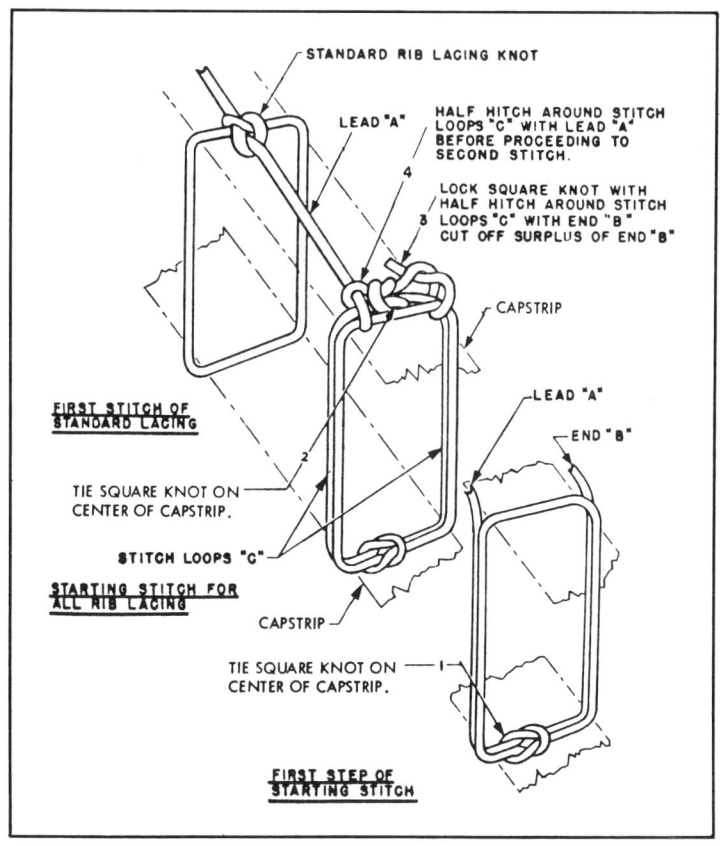

Fig. 7-9. Starting stitch for rib stitching.

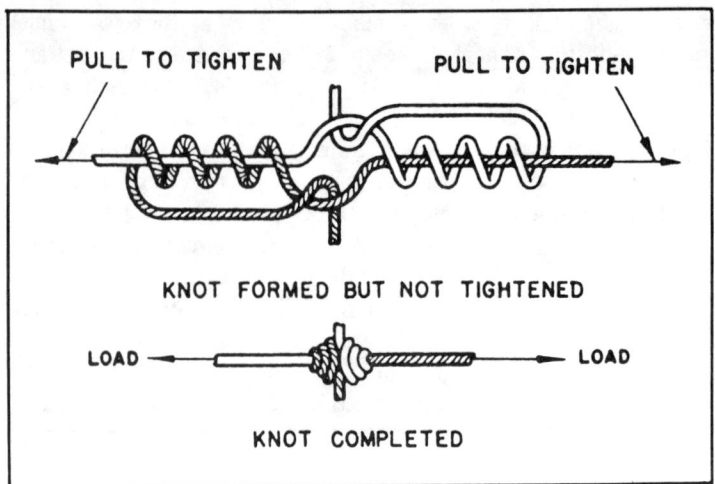

Fig. 7-10. Splice knot.

"wool" setting), shrinks to a satisfactory degree of tautness—such as produced by approximately five coats of dope on a cotton fabric. Discontinue the heat process when the envelope has acquired the approximate desired degree of tautness. Do not over-tauten. On a strong airframe such as a DC-3, B-25, etc., an increase in heat will not cause airframe distortion. However, with light airframes such as sailplanes, Aeroncas, and Cubs, care must be used not to over-tauten. Since Ceconite fabric shrinks immediately, only two seconds of heat application is required. Longer application of heat does not generally produce further shrinkage. Furthermore, damage to wooden members, other fabrics and electrical wiring could result from prolonged (10 seconds) application of heat above 248°F. On shrinking large surfaces (i.e., fuselage and wings), best results are obtained using two or three applications of heat, removing the slack on the initial go around with the main tautening on the second coverage. This is similar to spray painting, using two coats rather than a single massive coat. For initial shrinkage, it is preferrable to utilize a two man team, each equipped with a steaming steam iron or an electric iron set on "wool" and working on opposite sides. In rare cases where satisfactory tautness is not produced by this degree of heat, the temperature may be raised. Temperature steps of 25°F should be applied—up to 400°F maximum—until desired results are obtained. Wrinkles in Ceconite 101 fabric will vanish as the fabric is heat tautened. Since Ceconite is heat shrunk (and heat set) at a temperature higher than would be attained by leaving the airplane exposed to desert sun, no further shrinkage will occur by leaving the airplane in the hot sun.

For the tautening process, a hand-held source of controlled heat is required and the electric iron method is the preferred method. A household steaming steam iron is an excellent and safe source of controlled heat for removing general slackness. An electric iron set on "wool" temperature should be used in direct contact with Ceconite for final tautening. Due to the low heat output, electric heat lamps of less than 500 watt ratings are generally slow and unsatisfactory.

Whichever heat source is employed, best results are obtained by keeping the heat source in motion at a rate of approximately 5" per second. A motion akin to ironing is most satisfactory. First, side to side, then covering the same area in a top to bottom ironing motion to insure uniformly shrinking all areas.

Care must be taken not to heat this fabric in excess of 450°F as the fabric melts and strength decreases seriously. If through error this fabric heated above 450°F, the destroyed area will lose its cloth weave and assume a clearly deteriorated celluloid-like texture.

Small areas of doped Ceconite may be further tautened with heat by cautiously applying an electric iron in direct contact at a slightly increased temperature setting. However, prolonged application of heat (10 seconds or more) at temperatures of 300°F or more may cause discoloration of dope. Caution must be stressed. This step should be eliminated if possible.

Non-shrinking nitrate and butyrate dopes will generally not produce further measurable tautness. However, regular butyrate dopes will continue to tauten over a period of months and on light framed airplanes the envelope must be somewhat looser to allow for further shrinkage when using butyrate dope. As a rule of thumb, maximum tautening is achieved if a coin will bounce when dropped on the fabric surface. If normal butyrate is used, do not accomplish maximum tautening.

9. For covering plywood surfaces, Ceconite 103 is preferred. Specifically, plywood surfaces will be first refinished and sealed. Any irregularities will be filled and smoothed to eliminate condensation formation which could result in wood rot. Next, apply four coats of aircraft dope (nitrate butyrate) to the plywood surface. After the dope dries, attach Ceconite 103 to the plywood surface using "Super Seam" or aviation cement in a 1" wide strip around the edges. When dry, shrink Ceconite 103 with a steaming iron, or household iron in direct contact, to a glove fit. Finally, brush on two

coats of thinned clear dope and complete the finish (see Ceconite Appendix A).

Entrapped air causing bubbles must be removed. In general the finished surface must have a smooth appearance with no obvious airflow impingements (Figs. 7-11 and 7-12).

10. When Ceconite 102 and Ceconite 103 are applied, the method of attachment, repair and finish will be the same as that prescribed for Ceconite 101. When Ceconite 103 is used as a covering for gliders and sailplanes the method of attachment, repair and finish will be the same for Ceconite 101.

Repairs

11. With FAA Advisory Circular 43.13-1A and 2 (Section 3) as a guide, repairs are effected using Ceconite 101 and D-207 hand sewing thread. "Super Seam" or aviation cement is used for the adhesive and the procedures previously described are utilized. For small patches it is not necessary to remove the finish coat. Merely scuff the surface with No. 320 sandpaper, use "Super Seam Cement" and finish in a normal manner. Small patches of Ceconite do not require heat tautening.

When making repairs it is important to remember that nitrate dope will not adhere to butyrate dope because the nitrate solvents are of a lower order and cannot dissolve butyrate dope.

Annual Inspection

Aircraft covered in accordance with this process should undergo fabric inspection as prescribed in FAA Advisory Circular 43.13-1A and 2 for fabric-covered aircraft. Under normal care our tests indicated that an airplane covered with Ceconite fabric has three times the life of an airplane covered with conventional cotton fabric. With proper care, this can be the life span of the airplane. Although punch type testers are used with Ceconite it is generally felt that the ravel strip method is more satisfactory.

Note A: Under no circumstances will substitutes be made for structural items such as Ceconite fabric, D-693 or D-415, D-207, and D-69 cords and threads.

Note B: Extreme tautness can usually be relieved as follows: 1) Where "Super Seam Cement" attach methods are used, acetone or methyl ethyl ketone (MEK) may be used to soften the "Super Seam Cement" laps to relieve excessive fabric tautness. MEK and acetone will quickly evaporate with no impairment to the bond. 2)

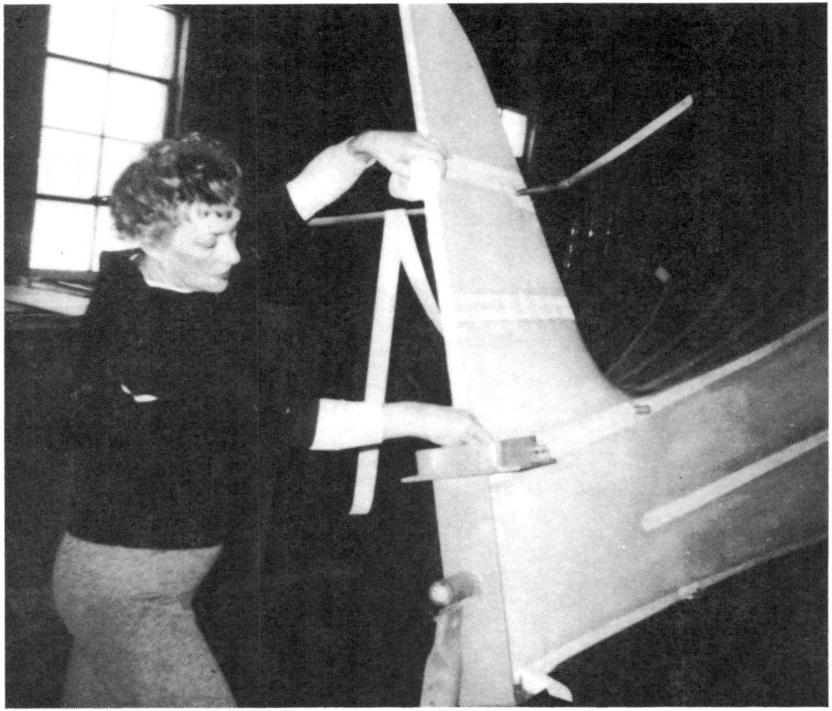

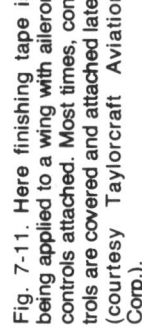

Fig. 7-11. Here finishing tape is being applied to a wing with aileron controls attached. Most times, controls are covered and attached later (courtesy Taylorcraft Aviation Corp.).

Clear lacquer or acrylic enamels combine with and dilute the dope solids to a degree which often relieves extreme tautness.

Note C: "Super Seam Cement" Bond—for maximum strength, care must be exercised to obtain good bonds. First, apply "Super Seam Cement" liberally to both surfaces. (Sufficient "Super Seam Cement" must be used to penetrate and completely encase each fiber.) Then, using your hands, join the surfaces as the cement becomes tacky. Finally, go over the outside of the joint using thinned cement. (Thin cement 10% with acetone or methyl ethyl ketone.) Hands may be cleaned with acetone.

Note D: Control surface CG determination: 1) Before removing old fabric, accurately weigh and determine CG of each control surface. Record this data and retain for possible FAA inspection usage. After recovering and refinishing again, reweigh and determine the CG. This step is important on high speed airplanes. 2) An easy method of recording CG data is to suspend the covered control surface by its hinges and attach an empty water pail in accordance with diagram. (See Fig. 7-3). Add water until the control surface balances in level position. Next, weigh the pail of water and record data. When the control surface is refinished, repeat this process and compare readings. If the final pail of water weight is the same or less than original weight, the surface is acceptable. However, if the final pail of water weight exceeds the original weight, the surface must be rebalanced in accordance with manufacturer's instructions.

Ceconite Appendix A: Standard Finish

Experience gained to date indicates the following finishing method to me most satisfactory in addition to effecting savings in labor and dope. The resulting finish is smooth and hides most of the fabric weave.

Step 1: Three brush coats of nitrate dope with the initial coat thinned about 30% with nitrate thinner and the other coats thinned to brushing consistency. A quart of "Super Seam Cement" added to each gallon, particularly the first coat, improves adhesion.

Step 2: Accomplish rib stitching, lay tapes and install grommets (use plenty of dope). Here also a quart of "Super Seam Cement" per gallon of dope will improve adhesion.

Step 3: For the buildup coats, spray or brush three coats of clear dope and then two coats of aluminum dope, using 2 to 4 oz. of aluminum powder per gallon before thinning. (Use of more aluminum powder per gallon will probably result in peeling.) Aluminum dope is used to insure against ultraviolet attack of fabric.

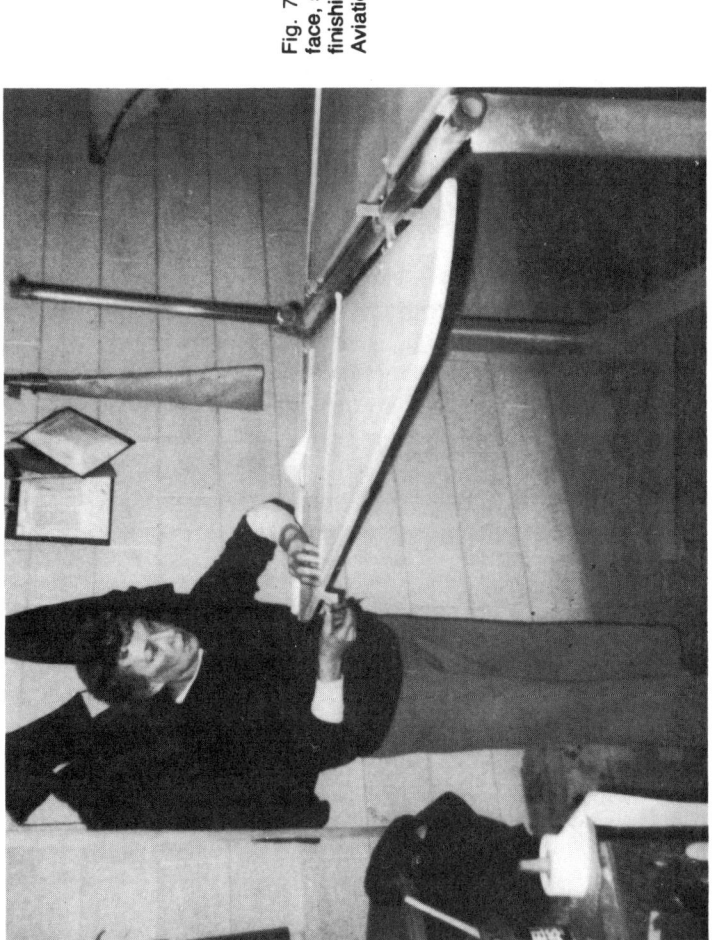

Fig. 7-12. An elevator control surface, shown on rotating stand, gets finishing tapes (courtesy Taylorcraft Aviation Corp.).

Step 4: Very light sanding with No. 320 Wet-or-Dry sandpaper, primarily to remove cotton nap (if cotton surface tapes are used) and any blemishes. Ceconite is a smooth multifilament and requires but little or no sanding at best. At this time, if the surface is opaque to light you may proceed to the color coats, otherwise an additional aluminum coat is required.

Step 5: For a good finish three color coats are generally required. Somewhat better gloss will result if the final color has approximately 30% retarder added.

Important: Be certain your material is FAA approved. All Ceconite 101 material must be stamped "FAA PMA CECONITE 101" at one yard intervals along the selvage edge and also have gray or black identification lines running the length of the material.

Note 1: It is necessary to achieve dope penetration of fabric and thoroughly coat Ceconite fiber on all sides in order to achieve mechanical adhesion in addition to the natural adhesion of dope. Nitrate dope is preferred as having better adhesive qualities than butyrate dope, and is preferred for Step 1 even if the final finish is to be butyrate dope. (Butyrate dope may be used after Step 1 if desired.)

Note 2: While many prefer enamel for the color coats, this procedure is not always advisable because of poor enamel-to-dope bond and the fact that a rejuvenation or major repair requires removal of the enamel. Furthermore, when enamel is applied over non-tautening butyrate dope (and probably over non-tautening nitrate dope also), the plasticizers may migrate, preventing cure of the enamel. If an enamel finish is mandatory be sure to thoroughly sand the dope surface to provide for a mechanical bond. Also avoid the use of non-tautening dopes. For Arctic service, the all-dope finish is by far the best.

Note 3: For gliders and sailplanes using Ceconite 103, experience has shown that all coats can be sprayed rather than brushing the initial coats since brushing tends to produce drop through. Furthermore, the number of clear coats may be reduced by 50%.

Note 4: Butyrate dope is somewhat more fire retardant than nitrate dope and more durable for a dope finish. However, butyrate dope, contrasted to nitrate dopes, continue to tauten over a period of months and have slightly more tautening effect. While these dopes have but little overall tautening effect on Ceconite, care should be exercised not to have the covering too taut when using butyrate dope.

Ceconite Appendix B: Alternative Finish

As an alternative to the finishing procedures previously mentioned, the Dac-Proofer® and Spra-Fill® system has also been found to be extremely acceptable for use when finishing Ceconite fabric. The procedure used in utilizing these products is as follows:

Step 1: Two (2) brush coats of Dac-Proofer®. No thinners or additives are necessary.

Step 2: Accomplish rib stitching, lay tapes, and install grommets (use plenty of dope). A quart of "Super Seam Cement" per gallon of dope will improve adhesion.

Step 3: For the buildup coats, spray three cross-coats of Spra-Fill®. Thin only as necessary for spraying. Spra-Fill® contains aluminum to protect the fabric against ultraviolet light.

Step 4: Very light sanding with No. 320 Wet-or-Dry sandpaper, primarily to remove cotton nap (if cotton surface tapes are used) and any blemishes. Ceconite is a smooth multi-filament and requires but little or no sanding at best. At this time if the surface is opaque to light, you may proceed to the color coats, otherwise an additional coat of Spra-Fill® is required.

Step 5: For a good finish three color coats are generally required. For a high gloss finish, a final spray coat of Clear CAB dope to which 10% color has been added, reduced equal parts with retarder and sprayed over the entire job will eliminate the need for rubbing or polishing.

Important: Be certain your material is FAA Approved. All Ceconite 101 material must be stamped "FAA PMA CECONITE 101" at one (1) yard intervals along the selvage edge and also have grey or black identification lines running the length of the material.

Note 1: It is necessary to achieve dope penetration of fabric and thoroughly coat Ceconite fiber on all sides in order to achieve mechanical adhesion in addition to the natural adhesion of dope. Dac-Proofer® is preferred as having better adhesive qualities than nitrate or butyrate dope, and is preferred for Step 1 even if the final finish is to be butyrate dope. (Butyrate dope may be used after Step 1 if desired.)

Note 2: While many prefer enamel for the color coats this procedure is not always advisable because of poor enamel-to-dope bond and the fact that a rejuvenation or major repair requires removal of the enamel. Furthermore, when enamel is applied over non-tautening Spra-Fill® and probably over non-tautening nitrate dope also, the plasticizers may migrate, preventing cure of the enamel. If an enamel finish is mandatory, be sure to thoroughly sand the Spra-Fill® surface to provide for a mechanical bond. For Arctic service the all dope finish is by far the best.

Note 3: For gliders and sailplanes using Ceconite 103, experience has shown that all coats can be sprayed rather than brushing the initial coats since brushing tends to produce drop through. Furthermore, the number of Dac-Proofer® coats may be reduced by 50%.

Ceconite Appendix C: Alternative Finish

The new Ceconite 7600 Finish is now approved by the Federal Aviation Administration under SA1351WE and SA2666WE for the use on both heavy and medium weight Ceconite fabric. There are some distinct advantages in the 7600 finishing process as compared to utilizing nitrate and butyrate dope with Ceconite 101 and 102. The advantages will be discussed and better understood after the 7600 procedure is explained.

Step 1: In using the 7600 series, it is necessary to utilize the precoated Ceconite fabric 76102 (pre-coated Ceconite 102) or 76101 (pre-coated Ceconite 101).

Step 2: 76102 or 76101 fabric is applied to the airframe utilizing either the blanket or envelope method. The method of tautening is identical to the previously mentioned method found in this procedure manual.

Step 3: 76102 and 76101 fabrics are yellow in color. When excessive heat is applied, this color temporarily changes to a light brown. This characteristic should serve as a warning signal to applicator. Ceconite reinforcing tape, machine and hand sewing threads (D-97 and D-207) and rib cord (D-693) is utilized as it is when recovering with Ceconite 102 and 101. 7602 cement and 7603 cement activator is to be used with the blanket method and when applying surface tape. Cement laps of three inches minimum width must be made when using cement to attach fabric to the structure, allow at least one hour before heat tautening the fabric. Waiting time can be shortened by using the tautening iron adjusted to medium heat after the cement is dried to the touch. Surface tape (7624) is applied for attachments and single stitch seams. Reinforcement patches are applied at all points where inspection openings are to be cut. Reinforced patches are to be cut from either 76101 or 76102, using 7602 cement and 7603 activator. After tapes and other reinforcements have been installed, one coat of 7603 activator should be applied over the surface of these items. Surface tapes need only to be installed over rib stitches or other attachments and over sewn seams. It is not necessary to apply surface tapes over seams that are cemented to the structure, however, they may be desirable from an appearance standpoint.

Step 4: The aircraft now has the fabric applied, heat tautened with reinforcing tape, rib stitch and surface tapes applied. The first finish step involves the use of 7601 filler coat. It is not necessary to brush the filler coat into the yellow pre-coated fabric. A chemical reaction will occur, resulting in a chemical bond between 7601 filler coat and 76101 or 76102 fabric. Before using 7601 filler coat, make sure it is mixed thoroughly. If material has settled, the use of a kitchen blender has been found quite satisfactory. For pad application, pour the amount of coating that will be used into a container slightly larger than the applicator pad. Dip the pad into the coating until completely saturated. Holding the pad flat against the fabric, apply one full even coat. One side of a typical wing can be coated by this method in about 20 minutes. After coating has dried, a second full wet coat is applied. This should be applied by wiping the pad across the surface at a right angle to the direction used in the first application. Drying time for each coat will range from 10 minutes to one hour depending on temperature, air flow, and humidity. When using spray equipment, coating may be applied in one full wet coat or two thin coats. The first coat spanwise on a wing and the second coat chordwise. The two coat system will result in less tendency for the coating to run or sag. Sanding: After the filler coat has hardened sufficiently (15 minutes to one hour depending on drying conditions) it is sanded. Just enough sanding should be done to remove any lumps or unevenness due to rough tape edges or dirt that may have gotten into the wet coating. This operation should require no more than ten minutes to sand one side of the average wing. If the surface is smooth and free of foreign matter, this operation may be omitted. Excessive sanding will reduce the coating thickness. Dry sanding is recommended. The best results are obtained using a 150-C-No-Fil Durite paper or equivalent. Do not rub in one spot for a long period of time as the friction will heat the coating and cause it to load the paper. If desired, the coating may now be sanded with either a wet or dry abrasive up to and including a 400 grit. Following this sanding, the surface should be thoroughly washed with water and allowed to dry. No aromatic or ketone type solvents should be used to clean the filler coating as they will soften or dissolve it. Most shops have found that a high speed orbital type sander is a good investment because of the sanding time savings and the very smooth finish obtained. If, by accident, the coating is sanded through to a point where the nap of the fabric starts to raise, filler coating may be reapplied to these areas and sanded.

Step 5: After the 7601 filler coat has been sanded as recommended, the following finish systems have been found satisfactory: 1) 7630 series pigmented urethane enamel; one fog coat, let dry 15 to 30 minutes, then follow with a wet coat. 2) Two cross coats pigmented butyrate dope. 3) One cross coat pigmented butyrate dope. Let dry two hours minimum. One fog coat Eonnex 7630 clear urethane enamel, followed in 15 to 30 minutes with a full wet coat. 4) One coat 7640 pigmented emulsion. Let dry. One fog coat Eonnex 7630 clear urethane enamel, followed in 15 to 30 minutes with a full wet coat.

7630 urethane enamel is an extremely flexible high gloss material. As a top coat for Eonnex cover, it provides crack and abrasion resistance even after a long aging. It is superior to anything else tested. It may be used over properly primed metal. This enamel is a two part mix. The thinners used are flammable. Adequate ventilation should be provided.

7640 pigmented emulsion paint is available in a wide range of colors. It is applied over Eonnex 7601 filler and properly primed metal. Only one coat is required, enough to provide the desired color. This coating is not designed to provide a long term service life unless it is protected by a clear overcoat, which provides the gloss and wear surface. 7640 emulsion paint offers advantages not available with any other FAA approved aircraft covering system. Multi-color paint jobs can be accomplished in a minimum of elapsed time. Drying time is very fast, five minutes to one hour, depending on ambient temperature and humidity. Masking for other colors can be done as soon as paint is dry. Sags or overspray can be washed off immediately with a wet cloth. The paint is thinned with water; however, it is not water soluble once it has dried. Overspray does not carry great distances, therefore, less masking is required. Fire hazard is zero and there are no toxic fumes.

Always test masking tape with 7640 paint before use. Emulsion type paints will frequently bleed under or soften conventional crepe back tapes. Film back and flat back waterproof tapes are superior. There is a difference between brands.

7640 paint may be recoated as soon as it is dry to the touch. It may be wet sanded when it is thoroughly dried to the extent that it cannot be scraped off or lifted by your fingernail. The paint should also be allowed to thoroughly dry (preferably overnight) before applying 7630 urethane enamel to insure that it is free of moisture that would react with the urethane. The emulsion paint may also be overcoated with clear butyrate dope. However, this is not a very durable finish. Do not stack urethane finished surfaces until they have thoroughly dried. Should anything stick to the surface, it will mar the gloss. Urethane finish should be applied when the temperature is 70°F or higher, low temperatures substantially lengthen the time to dry. Excessively high humidity is undesirable.

Helpful Hints

Filler, color and glaze application:

1. It is recommended that the 7601 filler coat be sanded until smooth and just enough remains on the fabric to block most light passage and fill the weave, (about .0083" thickness using 76101 fabric). Under some conditions, the filler coating may not flow out evenly when applied to the fabric and pinholes may be visible. Wetting the fabric with a water wet sponge just prior to a thin coating application will eliminate pinholing and improve flow out. After drying, this should be followed by a thin cross coat, then lightly wet sanded. This system helps provide a smooth finish and improves the hiding of the fabric weave. It may also be used with 7640 color coating.

2. 7640 color coat should be thinned with water only to the extent that it will not curtain when applied to vertical surfaces. Always thoroughly shake or stir the coating in the container and strain it when pouring. White color coat will hide 7601 filler when applied to a film thickness of only .0007". Theoretical coverage of 800 square feet per gallon. Deep tone and ultra deep tone colors may not hide as well. It may be desirable to apply a tint base before applying certain ultra deep tones. 7640 color coat is a very high solids material. It is expected that about six quarts will coat an entire typical light aircraft.

3. The filler and color coatings are softened and swelled to varying degrees by certain solvents, such as ketones, aromatics and some alcohols. Gasoline and oil have little or no effect on them.

4. The dry color coat may be given a light spray coat of 7631 reducer. After the solvent has evaporated (overnight) it may then be sanded to a smooth surface. This optional treatment tends to reduce the amount of fabric weave visible after the 7630 glaze coat has been applied. Wet sanding cuts the fastest.

5. Excessive filler and color coat thickness is not desirable from the cost and weight standpoint. It may also cause the coatings

to develop very slightly raised bumps resembling "measles" when a heavy glaze coat is applied.

6. Eonnex 7630 Glaze Coat is a very high solid material, about 45% by volume. Theoretical coverage is about 720 square feet per gallon per mil dry thickness. Standard recommended dry finish is one to two mils. The coating is mixed by volume, one part 7630A, one part 7630B, one part 7631 reducer. Stir well and strain. Application properties will be improved if the mixture is allowed to age thirty minutes before use. Additional reducer may be required in very hot and humid weather or when nearing the end of pot life of the mixed coating. Approximate dry times (hours): dust free, one; print free, three; dry hard, eight, at 78°F and 50% relative humidity.

Paint spray equipment that will handle conventional lacquers or enamels can be used to apply 7630 Glaze Coat. The thinners used are aromatic. Equipment clean up can be accomplished using lacquer thinner, toluene, M.E.K., or 7630 reducer. Do not allow mixed coating to remain in the equipment for more than 8 hours.

Advantages of 7600 Finish

1. Fabric to finish adhesion far superior to adhesion of nitrate dope to Ceconite.

2. Finish remains flexible and does not become brittle with age due to plasticizer migration.

3. 7600 is less expensive and easier to apply than any system currently on the market.

4. 7600 will not support combustion.

5. 7600 can be applied under almost any temperature and humidity conditions (except final finish coat).

6. 7600 provides a substantial weight savings.

7. 7600 is solvent free and therefore will not produce toxic fumes (except final finish coat).

Based on the above, Ceconite Inc. recommends the use of 7600 finish along with 76101 and 76102 fabric on all fabric aircraft.

By carefully following the instructions, you will obtain as nice a finish with synthetic fabrics as is obtainable with the cottons and linens. Also, the use of the newer synthetic fabrics, such as Ceconite 76101 and 76102 pre-coated fabrics, make the job easier from the standpoint of precise heat shrinking. They are considerably more error free, since they change color if too much heat is applied during the shrinking process.

Reassembling the Aircraft

When you have completed the entire process of applying the fabric, doping and painting the fuselage, wings and control surfaces, you are in the exciting, though time-consuming phase of reassembling the aircraft. I say exciting, because now, you can see the light at the end of the tunnel on the project—the airplane will eventually fly.

At this point, I would strongly suggest that you get your mechanic supervisor to assist you with the entire reassembly process. The reasons for this are that the reassembly and rigging of wings and controls is a job that a qualified mechanic should be involved with, and because the mechanic will have to make a final inspection before signing off the entire job as airworthy. You will help, of course, but rigging is an **exacting process** with little room for error. Consult the service manual for your airplane for details as to the precise rigging instructions. Consult also AC 43.13.

Once the painting is complete and dry, reattach the landing gear and hook up the brakes. Then move on to attaching and rigging the wings and controls. When all the wings and control surfaces are properly attached, rigged and safetied, you and your mechanic will reinstall the engine and all controls, wires and baffles. He will then do a differential compression check, which will be noted in the engine log along with the entry for the reinstallation of the engine and returning it to service. Then the cowling which has been repaired, where necessary, and repainted, can be installed. Then comes the interior.

The easy way to redo an interior is with pre-sewn headliners, available from many aircraft suppliers, such as Airtex Products. (See Sources of Materials and Supplies for names and addresses of suppliers.) The headliner you purchase will be accompanied by detailed instructions on installation. It is a process that will take two people the better part of a day to complete. The cost of a complete headliner is usually around $40 for most light two to four-place aircraft.

Reupholstery of the cockpit side panels can be accomplished either through your own covering of the original side panels, or the purchase of a side panel kit. With this finished, you will have put the final finishing touches on your restoration project.

There are some final items between you and the first flight, such as finishing the paper work. To do this, the airplane will be re-weighed, and an FAA form 337 completed to show the new weight and balance. This your mechanic will do, using airplane scales and a level. When the weighing and sign-off are completed you're ready to fly.

In the Air Again

If you have spent the four months to a year that it usually takes to finish a recovering project, and haven't flown in the meantime, by all means, get an instructor to accompany you and help you to get current. Too much work has gone into this project to spoil it through lack of proficiency.

But the moment will arrive when you taxi this enormous amount of sweat and toil out to the runway and take off. It is the single biggest moment of the whole project, as well as the beginning of years of enjoyment from your like-new airplane. You will now be flying your own inexpensive airplane and along with the pride of ownership will be the added pride of having rebuilt it entirely yourself. Added to the peace of mind that results from knowing that the job was done right, it all adds up to a feeling of tremendous satisfaction and accomplishment.

Chapter 8
Refurbishing Metal Aircraft

If you choose a metal airplane, there may be a lot of restoration work required on the metal surfaces of the outside and inside of the fuselage as well as the wings, control surfaces, struts, etc. Corrosion may be a problem in older, cheaper metal airplanes, especially those airplanes that have spent much time around coastal areas. Oxidation of metal surfaces and components is even more likely in the metal refurbishing project, and so the following, reprinted courtesy of Stits Aircraft Coatings, gives a good general description of one method of handling metal reconditioning and painting yourself. By following such instructions, you can save a very large percentage of the cost of restoring a metal aircraft, just the same as the fabric aircraft restorer saves time by installing fabric.

Thorough cleaning by Scotch-Brite scrubbing and a chromic acid conversion treatment is the recommended minimum preparation for a reasonable service life from any paint system on aluminum surfaces which will be exposed to the elements. Phosphoric acid etch before chromic acid treatment is recommended when severe environmental conditions are expected.

The procedures recommended here require coordinated work scheduling and seemingly repeated cleaning operations; however, they are considered the best to eliminate existing corrosion, provide lasting primer adhesion and protection against the paint film eroding and blistering in humid climates (filiform corrosion). Work scheduling on large projets with limited manpower or equipment is best handled by completing small sections or a few components at one time.

In most cases the decision on the painting preparation sequence is a compromise between the allowable painting budget and paint service requirements; however, a little more attention to details of the metal surface preparation can be the difference between paint problems developing in six months, or the paint remaining in excellent condition six to ten years, or longer.

The so called "Chromated" enamels used for economy purposes on aluminum without a conversion coating treatment or a primer will usually start peeling within a year. The use of non-corrosion inhibiting primers on top of an inadequately cleaned and treated surface is also a common cause of paint film loss or corrosion developing under the paint film in humid climates, indicated by blisters. Workmanship is 95% of a good paint job.

Aluminum Surface Painting

Old painted surfaces: Mask any area which will be damaged by paint stripper using cloth masking tape and polyethylene plastic sheet or equivalent material. Strip the old finish and thoroughly clean with Scotch-Brite abrasive pads or fine aluminum wool using water as a lubricant. Never use steel wool or emery cloth on aluminum surfaces due to corrosion from embedded microscopic particles. Any aluminum surface corrosion, indicated by discoloring and scaling, should be treated with Stits Aluma-Dyne E-2310 or E-2311 phosporic acid etch and brightener while scrubbing to show new metal. Phosphoric acid treatment of the entire surface after scrubbing is recommended and serves to provide microscopic tooth adhesion and a very clean surface for the primer bond. After etching, thoroughly rinse all acid from the surfaces with clean running water or clean rags rung out in clear water. Do not rinse with hot water. Air blow hinges, seams, or joints which may trap acid.

Bare oxidized surfaces: Surfaces heavily coated with oil, dirt, or mud should be cleaned with Stits XOFF-310 Alkaline Cleaner. After alkaline cleaning, mask any adjacent areas or openings which may be damaged by acid contact.

Old oxidized aluminum surfaces should be scrubbed with Scotch-Brite abrasive pads using diluted XOFF-310 alkaline cleaner as a lubricant, then treated with Aluma-Dyne E-2310 or E-2311 phosphoric acid etch. Pitted or corroded areas should be scrubbed while etching to show new metal. Thoroughly rinse and dry as instructed above.

New aluminum surfaces: Remove all dirt, light oxidation, oil, and residue of any protective coating by scrubbing with Scotch-

Brite abrasive pads using diluted XOFF-310 alkaline cleaner as a lubricant. Wash with clean water, dry, mask as needed, and treat with phosphoric acid etch as instructed above.

Treat all aluminum surface with Stits Aluma-Dyne E-2300 chromic acid conversion coating before the surfaces become contaminated or within eight hours after etching or Scotch-Brite scrubbing. Use a nylon brush or cellulose sponge and wear rubber gloves and a face shield. If more than eight hours elapse before chromic acid treatment, repeat the phosphoric etch or scrubbing operation. After the chromic acid treatment, wash with clean water and dry with clean rags. Air blow all water from hinges, joints, and seams or allow to evaporate. Chromic acid is not harmful if trapped in joints or seams.

Wipe all surfaces with a clean, untreated, lint-free, knit-free wiping or polishing cloth or similar paper wipe towels wetted with Stits C-2200 Metl-Sol cleaner just before priming. Shop towels furnished by commercial towel services are not recommended for final cleaning because they often contain silicones from previous polishing operations which are released by strong solvents and spread over the metal surface.

Apply two coats of Stits EP-420 corrosion resistant epoxy primer before the chromic acid treatment surfaces become contaminated and preferably within 24 hours after chromic acid treatment. Recommended primer dry film thickness is .06 to 1.0 mils. If the epoxy primer has aged more than three days before recoating, it should be scuff sanded with Scotch-Brite pads or 400 grit Wetordry (3M) sandpaper to remove the glaze and provide tooth adhesion. Avoid heavy scratches which will show through the high gloss finish coat.

The use of a pre-treatment wash primer (vinyl butyral-phosphoric acid resin) before application of the intermediate primer is recommended for the best protection and will improve the paint service life in very humid or rainy climates or on components subject to water submersion. Wash primer must be applied within the seven-hour elapsed time specified after mixing and recoated with the intermediate primer within 24 hours after application. Vinyl butyral wash primers are not considered to have sufficient pigment loading or film thickness to be used as a sole primer coat. An intermediate coat of conventional primer such as Epoxy EP-420 primer is recommended before the finished coat.

Apply the selected finish coat before the primed surfaces have become contaminated. Any inadvertent contamination such as fingerprints or oil may be safely removed with Stits C-2210 paint surface cleaner. Remove any dust or lint from the primer surface with a clean, good quality tack rag just before painting. Recommend three coats Stits Aluma-Thane Enamel at 15 to 20 minute intervals. Allow sufficient time between coats to avoid runs.

Many professional painters prefer to apply the lighter colors such as yellow, light blues, light reds, etc., over a white base coat to enhance the color shade. The white base coat should dry thoroughly before applying the color coat, to eliminate the possibility of the fresh white pigments migrating to the surface of the color coat.

Stits urethane enamels do not have the crawling and cratering characteristics inherent in many other brands of urethane enamels and are equal in quality to any brand on the market regardless of the price or advertising claims.

Recommended finish coat thickness of 1.3 to 2.0 mils., depending on the type of coating. The higher solids epoxy primer and urethane finishes will build a thicker film with each coat than the one part coatings and primers.

Steel Surface Painting

The steel surface must be free of all rust, oil, grease, tar, wax, and old paint scale to provide good adhesion and prevent corrosion developing under the primer. Excessively dirty and grimy surfaces should be thoroughly cleaned with equipment and methods according to the size and configuration of the object being refinished and cleaning methods available.

Light oil, grease, wax, silicone, and tar may be removed with Stits C-2200 Metl-Sol Cleaner. Heavy rusts, corrosion and paint scale should be removed with sand or glass bead blast.

Step 1: Working small areas, treat the previously cleaned surface with Aluma-Dyne E-2310 phosphoric acid etch solution, diluted equal parts with water or full strength depending on the surface condition. Use a nylon brush or cellulose sponge and wear rubber gloves and a face shield. Scrub with steel wood pads or wire brush during acid etch treatment to remove any light rust and scale. The acid solution should remain on the steel surface for five to ten minutes and then the surface is washed thoroughly with clean running water or wiped with clean rags rung out in clean water. Wipe dry with clean rags and air blow all corners, joints, and seams to remove any acid etch residue.

Step 2: Within a few hours after the etched steel surface is dried, apply two coats of EP-420 Corrosion Resistant Epoxy Primer. Do not allow the surface to become contaminated before primer application. Re-etch if necessary.

The use of a pre-treatment wash primer (vinyl butyral-phosphoric acid resin) before application of the intermediate primer is optional and will improve the paint service life in humid or rainy weather or on components subject to water submersion.

Step 3: After the primer has dried, apply the finish coats. For the most durable finish we recommend three coats of Aluma-Thane enamel.

Magnesium Surface Painting

Magnesium is the most chemically sensitive metal used in aircraft structures and must be protected from rapid oxidation after cleaning and conversion coating treatment through priming to assure a good primer bond.

The most common cause of primer and paint peeling from magnesium surfaces is inadequate cleaning, failure to touch up the existing conversion coating after damage during paint stripping operations, or delaying too long after cleaning and conversion coating treatment before primer application.

Strip the old finish and clean the surface with Scotch-Brite abrasive pads or fine aluminum wool using water as a lubricant. Do not use steel wool or emery cloth on magnesium surfaces due to corrosion from embedded particles.

If the surface is corroded or pitted, scrub with Scotch-Brite abrasive pads to show new metal before the conversion coating treatment.

Thoroughly clean the surface with XOFF-310 alkaline cleaner. Rinse with clear water and wipe dry with clean rags. Never use methyl alcohol (methanol) or wood alcohol to clean magnesium surfaces.

Apply Stits Magna-Dyne E-2390 with a nylon brush or cellulose sponge. Wear rubber gloves and face shield. Chemical action on fresh, untreated magnesium will be noted by foaming which will stop when solution is spent. Very little foaming action will be noted over old conversion coatings.

Keep the surface wet for one to three minutes. Continue wiping the surface to replenish the spent or drained off solution. One-minute treatment time produces a brassy iridescence. Three minutes produces a dark brown color, the best for paint adhesion.

Do not exceed three minutes because the result could be a loose, powdery coating.

While the surface is still wet with Magna-Dyne E-2390, rinse with clean water or wipe with clean, wet rags rung out in clean water. Do not use a hot water rinse. Air blow water from hinges, joints, and seams. Wipe dry with clean rags.

Wipe all surfaces with a clean, untreated, lint-free, knit-type wiping or polishing cloth or similar paper wipe towel dampened with Stits C-2200 Metl-Sol cleaner just before priming.

Apply 2 coats of Stits EP-420 Epoxy Primer within eight hours or before exposure to outside atmospheric conditions. Clean and repeat the conversion coating treatment if the surface remains exposed overnight.

The use of a pre-treatment wash primer (vinyl butyral-phosphoric acid resin) before application of the intermediate primer is recommended for the best protection and will improve the paint service life.

Apply three coats of Aluma-Thane Enamel at 15 to 20 minute intervals before the primed surfaces have become contaminated.

If EP-420 Epoxy Primer is used and has cured over three days before application of the finish coat, the primer surface should be lightly scuff sanded with Scotch-Brite pads or #400 grit Wetordry (3M) sandpaper to remove the surface glaze and provide tooth adhesion. Avoid heavy scratches which will show through the high gloss finish coat. Do not allow the primer surface to become contaminated by handling or overspray from adjacent work operations before finishing. Any contamination of the primer surface may be removed with C-2210 Paint Surface Cleaner.

Treating Oxidized or Corroded Aluminum

Surfaces heavily coated with oil, dirt, or mud should be cleaned with XOFF-310 alkaline cleaner. After alkaline cleaning, mask any adjacent areas or openings which may be damaged by acid contact, using cloth masking tape and polyethylene sheeting.

Working a limited area to provide adequate attention and time control, apply Aluma-Dyne E-2311 Phosphoric Acid Cream or E-2310 Phosphoric Acid Solution with a nylon brush or cellulose sponge. Wear rubber gloves and a face shield. Excessively oxidized, stained, or corroded surfaces should be scrubbed with Scotch-Brite Ultra-fine cleaning pads while etching to show new metal. Never use steel wool or emery cloth on aluminum surfaces due to corrosion promotion from embedded particles.

Thoroughly rinse with cool running water or wipe with wet rags or sponge rung out in clean water. Don't rinse with hot water. Air blow hinges, seams, or joints which may trap acid.

Allow surface to dry or wipe dry with clean rags and proceed with chromic acid treatment before the surface becomes contaminated or maximum eight hours after phosphoric acid etch or scrubbing. Aluma-Dyne E-2300 chromic acid treatment is recommended for corrosion protection whether the surface will remain bare, be waxed, or painted with a clear finish.

When a bright polished finish is required, the surface should be buffed before, not after, the invisible Aluma-Dyne E-2300 treatment. Remove any buffing compound residue with XOFF-310 cleaner before Aluma-Dyne E-2300 treatment. Add one spoon XOFF-310 to 16 oz. diluted chromic acid solution to break the water surface tension and provide improved flow out on polished aluminum surfaces. Use the modified chromic acid within seven days or discard. Do not polish the surface with an abrasive after the E-2300 chromic acid treatment or the invisible protective coating will be removed.

Inaccessible Areas and Corrosion Protection

Interior surfaces which need corrosion protection from salt water such as floats, wings, or aft fuselage may be alkaline cleaned, chromic acid treated, and small cloth bags of potassium dichromate or sodium dichromate placed in strategic locations where water may pool. Secure to avoid interference with any controls. Replace as needed.

The diluted XOFF-310 alkaline cleaner is sprayed through inspection access holes and any other openings with an engine wash gun. Follow with clean water and allow to drain.

Dilute Aluma-Dyne E-2300 chromic acid solution at a ratio of one part to three parts clean water and spray the interior with an engine wash gun in the same manner. Wear rubber gloves and face mask. If the component can be removed to handle, slush by rotating. A plastic sheet over supporting structures forming a trough will contain the spilled and reusable solution. Complete rinsing of the interior after 10 to 15 minutes is optional because the chromic acid solution is harmless after it is spent. After the chromic acid solution is thoroughly drained, drying may be aided by an air blast through the interior.

Painting Interior Surfaces

Interior aluminum surfaces, in good condition, which are to be painted only for cosmetic purposes, may be cleaned with C-2200 Metl-Sol cleaner or XOFF-310 alkaline cleaner and primed. Interior surfaces which will be exposed to corrosive elements such as sea water, should be cleaned, scrubbed, and treated with Aluma-Dyne E-2300 chromic acid conversion coating before priming.

Finishing Fiberglass Components

Strip any old paint or primer film, thoroughly clean, and sand with 320 grit Wetordy (3M) sandpaper. New surfaces should be cleaned of any mold release coatings using an appropriate solvent, depending on the type of mold release used. Silicone types may be removed with C-2200 Metl-Sol cleaner. Sand thoroughly with 320 grit Wetordry (3M) sandpaper to remove the surface gloss and provide tooth adhesion.

Apply one good coat of Stits EP-420 Epoxy Primer and finish with three coats of Aluma-Thane enamel. If the epoxy primer has aged more than three days before recoating, scuff sand to remove the primer surface gloss.

Refinishing Old Paint

Stits Aero-Thane enamel may be used to refinish sound acrylic and nitrocellulose lacquer, epoxy paint, and the better quality synthetic enamel successfully, depending on their condition. The old surface must be thoroughly cleaned and sanded smooth. Finish a small component first to test suitability.

Primers and Finish Coats

There are many brands of "zinc chromate" primers on the market and it should be understood that zinc chromate is only a corrosion inhibiting dry pigment which must be incorporated in a suitable vehicle or resin to provide adhesion and bond the zinc chromate pigments to the metal surface for corrosion protection. One of the most common vehicles used is based on low-cost oil alkyd resins. ("Synthetic" enamel is also an oil alkyd resin.) The better quality alkyd primers manufactured to Mil-P-8585 specifications provide good service on most commercial applications; however, many "economy" brands of alkyd primers are not compatible with the solvents in our pigmented finishes and may "lift." We do not recommend any of our coating products be used over alkyd type primers of any other manufacturer's brand due to possible incompatibility.

Epoxy primers are considered superior to all other types of primers and provide better adhesion to aluminum surfaces than do urethane primers, especially in water soak tests. Urethane primers do provide slightly more chemical and solvent resistance than epoxy primers; however, since the primer is coated with a suitable finish coat, it is not exposed to solvents and chemicals.

Good quality urethane finishes provide about twice the service life as epoxy finishes. Epoxy finishes are usually about as durable as synthetic enamel finishes and will chalk in about the same length of time; however, epoxy finishes provide far better chemical and solvent resistance than synthetic enamel finishes. It is our opinion, based on tests which correspond with many tests conducted by major material suppliers in the U.S. and Germany, that the best paint system for metal aircraft is epoxy primers and urethane finish coats.

Spraying Equipment

Spraying equipment which is rated for application of lacquer, nitrate or butyrate dope, synthetic enamel and shellac may be used to apply all Stits coatings.

A DeVilbiss Model #MBC-510 or JGA-501 spray gun with #30 air cap and EX tip and needle at 50 pounds pressure has been tested satisfactorily. Any alternate brand with equivalent air cap and needle size, either pressure pot or suction type will also provide satisfactory service. Correct adjustment of air-to-liquid ratio is important. Be certain the spray gun air requirement does not exceed the air supply or hose capacity.

Pressure pot painting equipment may apply up to twice the amount of material as the suction type gun on each coat. This should be taken into consideration when counting the number of coats applied to a surface. Ten pounds of air at the pressure pot with 50 pounds of air at the gun will apply about the same amount in two coats as the suction gun will in three coats.

Due to possibility of eye damage, special attention should be given to protecting the eyes from splashing liquids when working with any coating or chemical. A plastic eye shield is recommended. A face filter mask is also recommended when spray painting to protect the lungs from spray mist. Persons allergic to various solvent vapors should use an activated charcoal respirator mask or a hood with a fresh air supply from the air hose. Various type and brands are available from industrial supply stores. Exposure to light urethane enamel spray mist for as little as 30 minutes time will give many persons the symptoms of flu the next day.

Miscellaneous Information

Do not intermix or substitute reducers or thinners. Filter all coatings before application. Use proper spray gun adjustment, air pressure, and spraying technique. Follow all instructions for use of each product.

Do not try to "invent" new painting procedures, system combinations, or short cuts when painting a component that must be right the first time. Rework is expensive.

An adequate water trap should be installed in the air line to eliminate the possibility of moisture being mixed with the spray coat. Moisture will cause pinholes and blisters, which will show in the finish coat. Drain the trap often.

Do not try to spray paint in the following conditions: drafty, windy or dusty area; outdoors in direct sunlight if the weather is hot; outdoors in a heavy fog or when dew is forming; high humidity; below 60° or above 100°F.

Any grease or oil accidently deposited on the surface should be removed with a clean lint free rag dampened with C-2210 paint surface cleaner. Do not use engine cleaning solvent or gasoline because they leave a film.

Use only good quality solvent-resistant masking tape (3M #236 or equivalent) when taping trim lines to avoid solvent penetration through tapes and "seep" of trim colors under tape edges.

Chapter 9
Accessories

Once your bird is all back together and the paperwork is done, you have an airplane that is loads of fun to fly for the pure pleasure of being aloft. There is no end to the fun your airplane will bring you. But there are some additional activities you may want to consider as additional sport flying possibilities. These are floatplane flying and winter flying on skis. Also, some additional equipment to make your flying more versatile and enjoyable may be considered, as well as an exciting new alternative to conventional aircraft ownership on a very small budget: ultralights.

Floatplane Flying

Floats can be purchased through *Trade-A-Plane* and elsewhere for most of the older conventional gear singles such as the Piper Cub, Super Cub, Taylorcraft, Aeronca 7AC and Citabria. Cost can range from $1500 to around $3500 as a guideline. They are best suited to the 85 horsepower and up airplanes, but have been known to perform adequately with 65 ponies as well. Water flying opens up a whole new world of fun for the sport airplane pilot. The only requirements other than the floats is an additional rating on your pilot certificate.

You will need a five to ten hour course of seaplane instruction, after which you will take a flight test from the FAA or an FAA Designated Examiner. No written test is required, and when you finish, your license will proudly display the "Sea" after "Land" as in "Single Engine Land and Sea."

Winter Flying on Skis

As for flying off skis, this is a barrel of fun for those who live in or have reasonable access to the snow country. No licensing additions are required of the pilot, and only nominal changes are made to the airplane to add the skis and their rigging. Many ski attachments simply bolt onto the gear under the wheels, and as such require hardly any modification to the airplane (Fig. 9-1).

It is a good idea to get some instruction from an experienced ski pilot. Although most experienced ski pilots maintain that flying off the skis is actually easier than wheels, there are some peculiar considerations. One is the fact that you have no brakes, except for the drag of the tailwheel. Another is the fact that you can and probably will be making some landings off an airport, since in many ways, all the world is an airport to the skiplane pilot. Much care must be taken in choosing off-airport landing sights, in consideration of both safety and legality. So the best thing for the prospective skiplane fan to do is get together with those in your area who enjoy this sport and learn the ropes. Again, a little verbal flag waving will bring out the solutions to most problems and questions from among others in the "flying fraternity."

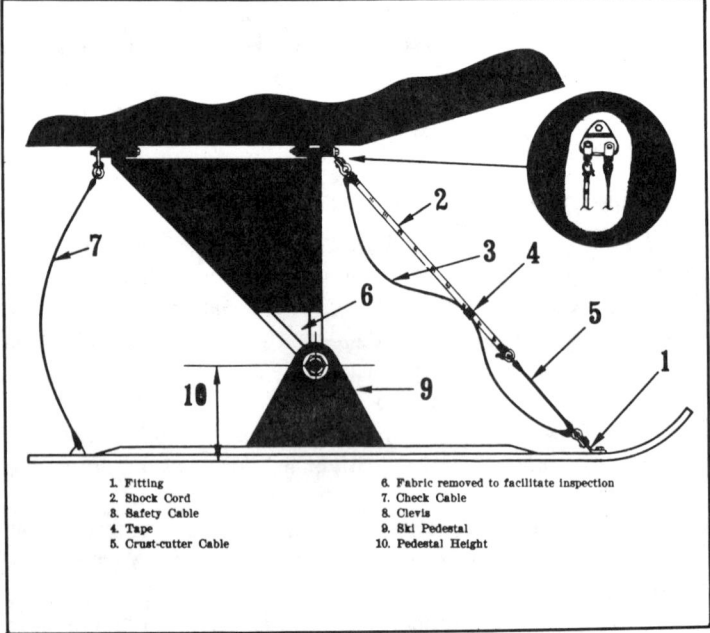

1. Fitting
2. Shock Cord
3. Safety Cable
4. Tape
5. Crust-cutter Cable
6. Fabric removed to facilitate inspection
7. Check Cable
8. Clevis
9. Ski Pedestal
10. Pedestal Height

Fig. 9-1. Typical ski installation.

Fig. 9-2. Hand-held VHF radios, such as this Terra TPX-10 might be the answer for light airplanes without electrical systems.

Electronics

Even if you do not keep your airplane at a controlled airport, you will probably sooner or later want to fly into and out of one, and you will need a two-way VHF radio for this purpose. In all probability, you bought an airplane without even an electrical system, much less a radio, since most of the light, older, single-engine airplanes built in the 1940s and 1950s were not equipped with electrical systems. You have basically two options open at a nominal price. One is to install an electrical system, and the other is to obtain any one of a number of small hand-held battery-powered VHF aircraft radios (Fig. 9-2).

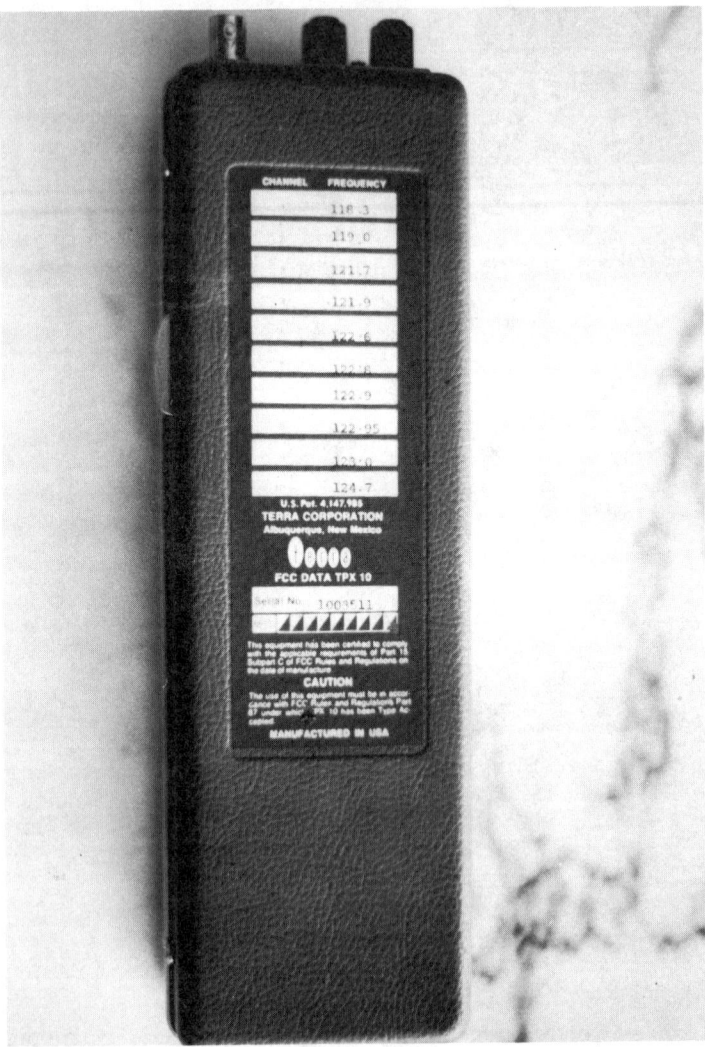

Fig. 9-3. Hand-held radios like the TPX-10 can be programmed for up to 10 frequencies of your choice, depending on the most available frequencies for your area.

Fig. 9-4. The frequency to be used is dialed into the channel selector on top of the set. Volume control, squelch and gain function controls are located for easy reach. Small mountings can be put in your airplane to hold the radio during flight; set can be taken home at night for safety.

Going the electrical system route is somewhat expensive in two important ways. One, you have to install the equipment, which usually consists of a wind-driven generator, powered by a small windmill attached to the belly of the airplane. This will produce enough electricity to operate a small radio, along with position lights and landing lights for night flying. It does, however, add weight. The weight of the system and radio can run to 30 pounds. In a 65 horsepower airplane, thirty pounds can make quite a difference. 85 horsepower airplanes do not suffer nearly as much. The advantage of an electrical system of this type is that it requires no heavy, expensive battery, and it will enable you to equip the aircraft with the necessary lights required for night operations. The disadvantages are in terms of cost and weight.

The other radio option is the hand-held VHF radio, which is becoming a popular addition to small airplanes, due to the low cost, low weight, and ease of handling.

An example of the typical hand-held, battery powered **VHF** aircraft radio, is the Terra TPX-10. It is small, weighs only a few pounds, and can be programmed with 10 channels or frequencies that you choose according to the frequencies most common to your area (Figs. 9-3, 9-4).

The cost of a radio such as the Terra TPX-10 is about $400 to $600. For most small airplane operations, it is a good idea to purchase the optional headset and boom microphone with push-to-talk feature to facilitate convenience and readability in the cockpit (Fig. 9-5).

Another advantage to this type of system is the fact that when you finish flying, you simply take the radio equipment home with you, thus precluding theft of a fixed or semi-fixed radio installation.

There are also numerous 720-channel hand-held **VHF** radios on the market, which though more expensive, offer the advantage of all the aircraft frequencies. In this way, the pilot is not limited to a few frequencies which may not be available in some areas to which he would like to fly. Cost of the 720-channel sets is normally around the $700 to $900 figure. It is doubtful, however, that any non-portable **VHF** radio, along with the required electrical system installation, could be added to an airplane for much less than $1200 to $1500.

Consult the listing of aircraft components suppliers in the back of this book (Sources of Materials and Supplies) for information about various radio equipment. Also, here is a list of some of the major suppliers of hand-held battery-powered **VHF** radios for light aircraft: 1) Terra Corporation, 3520 Pan American Freeway, N.E., Albuquerque, NM 87107, telephone: (505) 345-5621; 2) Exeter Avionics, 4828 Hollins Ferry Road, Baltimore, MD 21227; 3) Ritron, Candle Aviation Supply, Inc., 2111 Lakeland Avenue, Ronkonkoma, NY 11779, telephone: (516) 981-8880; 4) Aircraft Components, Inc., 700 North Shore Drive, Benton Harbor, MI 49002, telephone: (800) 253-0800.

One other tip to lightplane owners who have airplanes in which no radio has previously been installed: your engine probably has unshielded spark plugs that look a lot like regular automobile spark plugs. These will create a great deal of static in your radio. So it is a good idea to shop around for a good deal on a set of shielded spark plugs. Hang on to your old ones and either trade them in on the new ones or sell them to someone that uses that type. Aircraft spark plugs cost $4.00 to $7.00 each, so try to swap or sell yours if you can.

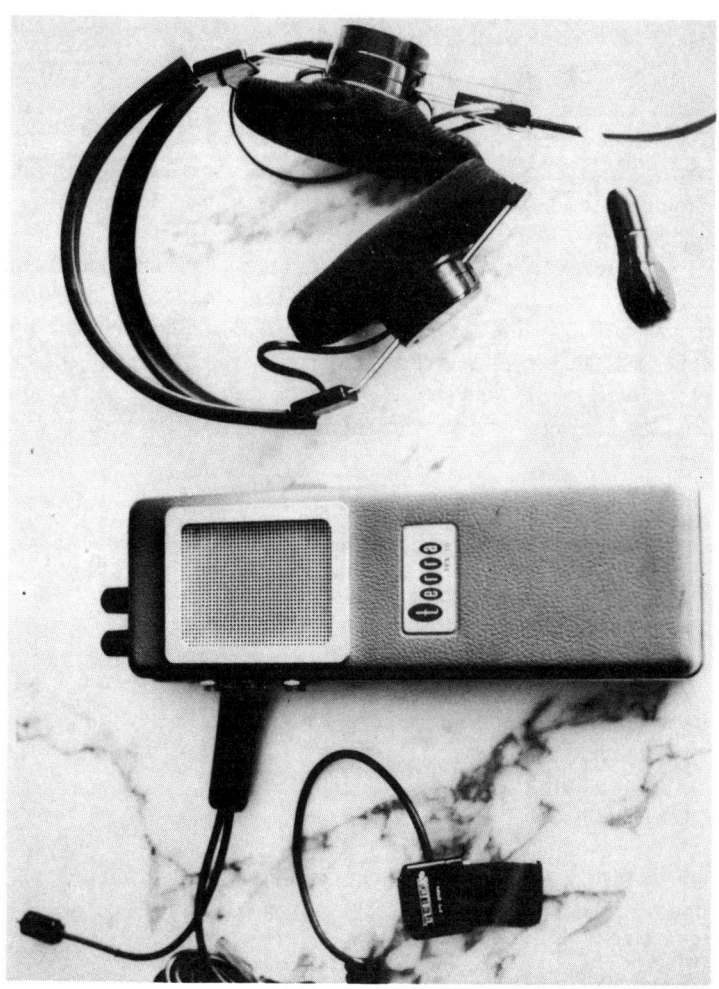

Fig. 9-5. Headset and boom microphones are really handy for hands-free radio communications.

Also on the subject of electronic equipment, be advised that the FAA currently requires that your aircraft be equipped with an ELT—or Emergency Locator Transmitter. This device sends out a signal if your airplane should ever crash. The units are not required for training flights conducted within 25 nautical miles of the home base of the aircraft.

You can obtain an inexpensive ELT from one of the aircraft supply stores if your airplane is not already equipped. If it has flown much in the last five years, however, it is probably already equipped with one. When I bought my last taildragger, I was not aware that it was equipped with an ELT, since the unit was buried in the wing root with only the tiniest of antennas protruding from the top of the wing.

The ELT broadcasts its signal on a frequency of 121.5 MHz, which is the standard aviation emergency frequency. One problem common to the ELT is that it has a bad habit of activating itself at the wrong times, such as during a solid landing, or sometimes in the middle of the night, sitting quietly in the hangar, for no reason whatsoever. I mention this just so you'll know.

Whatever type of electrical and/or other additional equipment you decide to install, remember that it will add both cost and weight to your airplane. It is easy to go overboard on add-ons and wind up with a $20,000 two-place airplane that has no useful load left to carry people and fuel. Remember that the object of this project is to obtain wings on a budget, and view the purchase of optional equipment in this light.

Ultralights: A Low Cost Alternative

For those who would like to fly for the pure love of being in the air, and would like flying a machine that is cheap to buy and cheap to operate, there is an alternative to conventional airplanes of the type that have been discussed here.

I am referring to the "ultralight" category of airplanes, such as the little Weedhopper shown in Fig. 9-6. They are small, inexpensive (around $3,000 new) very safe, and require no licensing, either for the airplane or the pilot, as of this writing, and you use your feet for a landing gear.

Almost anyone can fly them; they cruise at 30 to 35 miles per hour, and stall at 15 miles per hour. Landing speed is around 20 to 25 miles per hour. They are so easy to fly and land that people have been known to take off and fly them with but a few hours of ground instruction and no previous flying time! (Figs. 9-7, 9-8).

Fig. 9-6. The Weedhopper is a fabric-covered ultralight powered by a 25 horsepower Chotia 460-B engine. It is designed for one thing: fun! (courtesy Weedhopper of Utah, Inc.).

Fig. 9-7. Weedhoppers are kit-built funplanes that are inexpensive to buy and burn only one gallon of fuel per hour (courtesy Weedhopper of Utah, Inc.).

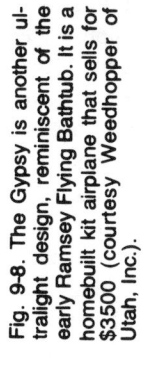

Fig. 9-8. The Gypsy is another ultralight design, reminiscent of the early Ramsey Flying Bathtub. It is a homebuilt kit airplane that sells for $3500 (courtesy Weedhopper of Utah, Inc.).

Another amazing fact about these diminutive ships is that their small engines burn only one gallon of fuel per hour.

Ultralights usually are sold in kit form, requiring the buyer to spend 20 to 40 man-hours labor putting them together according to detailed instructions. Anyone can do this, with only normal household hand tools. They are the essence of simplicity, and are really basic fun machines. I flew one myself, and although skeptical at first, really learned to love ultralight flying.

One of the leading makers of ultralight flying machines is Weedhopper of Utah, Inc., of Ogden, Utah. Contact them for more information on the entire field of ultralight airplanes: Weedhopper of Utah, Inc., Box 2253, 1148 Century Drive, Ogden, UT 84404, telephone: (801) 621-3941.

Chapter 10
Other Considerations

Rebuilding from Salvage

Although it has been repeatedly said here that it is wise to avoid the total "basket case" type of airplane as a choice for restoration, it is sometimes worthwhile to consider the possibility of restoring an airplane from the parts of two or more wrecked or otherwise "salvage" airplanes. Great care must be used by the non-mechanic when getting involved in a project from this point of view, however; it must be borne in mind that, in some cases, the work required to make a whole, airworthy airplane out of several salvaged ones may require the expertise and qualifications of an A&P mechanic. Sometimes, the cost of having a shop do the work that cannot be done by the amateur owner is prohibitive. So it is essential that a prospective restorer take a mechanic along when looking at wrecked airplanes to find out *exactly* what will need to be done to put the ship back in the air again.

If two airplanes are pooled to make one airworthy one, you must make sure that there are enough usable, airworthy parts to form a whole airplane. Frequently, you'll find one good wing on one and one on the other; a straight fuselage on one, a good tailwheel on the other. One may have a good engine, while the other will have no engine at all, but as long as the sum of all the parts will equal a whole airplane, the rest is hard work, and at the end, a lot of paperwork.

The paperwork comes in when any part or component that was part of an airplane is put on another airplane. This is particularly true when major airframe components, such as wings, fuselages, engines, etc. are transferred from one aircraft to another. The serial numbers of the parts, in the case of engines and powerplant components, and a complete description of the part and its method of application in the case of airframe components will be required for the airframe and/or engine logs of the resulting restored aircraft. Your mechanic will know what logbook entries to make.

As to the costs involved in restoring damaged airplanes, they can run the entire gamut from the bargain basement to the ridiculous. If you buy from an aircraft salvage yard, you will probably pay close to the retail price for all the serviceable components that are left on the damaged ship, since the salvage man will have gone through the airplane and inventoried what is and is not of value on the airplane. Often times, engines will have already been removed from wrecks, as will control surfaces, instruments, components, and so forth. The salvage airplane will probably not be an attractive-looking airplane when you view it in the yard, and the price may not mean much to you. A mechanic's advice in this situation is invaluable.

On the other hand, you will find many opportunities to purchase damaged airplanes directly from their owners. The airplane that has been groundlooped, nosed over, wind damaged, taxied into a ground obstacle, caught fire, etc., will in most cases be offered for sale by its woebegone owner. Here the bargaining and dickering might result in an excellent deal for you.

But the obvious problem is that if you find one damaged airplane in this way, you will still need another. This is partly true and partly untrue. If the airplane has been wind damaged and ground-looped, maybe the only damage was to the wings. You can repair these in the process of recovering, assuming we're talking about a fabric-covered airplane. In the case of metal airplanes, this type of damage is more difficult and much more expensive to repair. A wing, for example, on a popular airplane such as a Taylorcraft, Piper J-3, Aeronca, etc. can be purchased from companies such as Univair, a salvage company, or perhaps located via an ad in a national ad paper such as *Trade-A-Plane*. The same goes for propellers, engines, tailwheels, controls, and so on. Also, the more popular aircraft have parts available by virtue of many parts manufacturers who hold STCs or Standard Type Certificates for the manufacture of parts for particular aircraft. So if you are able to find a good bargain in a damaged, salvageable airplane, and you can determine that the pieces necessary to put the ship back in the air again are available from either another wrecked airplane or a commercial source, you may be on the track of a true bargain that will require only a large investment in time and effort to complete.

Engine Swapping

It is not uncommon to find a perfectly good, restorable airplane that has a "run-out" engine, i.e. one that is not airworthy because it is in need of a major overhaul. Frequently, the owner has neither the resources or the desire to have this done by an aircraft shop, so the airplane goes onto the sale block. An Aeronca 7AC, Piper J-3, any of the "PA" series Pacers, Tri-Pacers, Colts, etc., Stinsons, T-Crafts, rag-wing Cessnas, 120s, 140s, and so on, these aircraft will cost thousands less with run-out engines. This is where the engine swap can work to your advantage.

As stated earlier, the price of a factory-remanufactured or new engine in the horsepower range we've discussed here will be between $3500 and $5000. This is an "exchange" price, meaning that the purchaser trades his run-out engine for the remanufactured engine, plus the price. This is one alternative in the engine swapping method. Another is to watch the ads and find a low-to mid-time engine from an individual owner. Sometimes this will work fine, but there are risks. For one thing, you have no real way of knowing how the engine has been operated or cared for. Also, even the indication of a fresh overhaul in the logbooks does not assure you of *who* did the overhaul or *how well* it was done. The quality of workmanship varies greatly, even among certificated engine overhaul shops, so take great care when buying by this method, and, again, take a mechanic along to perform a differential compression check, etc., before ever putting out the money.

Beware of any engine offered for sale at a low price but "without logbooks." Logboks are the essential history of an engine, and if the logbooks are unavailable, there is no way to tell the maintenance history, the mechanical condition and performance, or for that matter, even the true age of the engine, unless it is marked on the nomenclature plate attached to the crankcase. Buying an engine that has such a dubious history would be a penny wise, pound foolish move for the pilot seeking to get into the air on a budget.

A factory-remanufactured engine, on the other hand, is warranted to be as good as new; indeed, it has been overhauled to factory-new tolerances, as opposed to the so-called "service tolerances" of the field major overhaul. Though more expensive, it is likely to give long and reliable service just like a brand new engine, and will increase the overall value of the aircraft. This method of replacing a run-out engine would be the best way, if the funds could be directed in this way.

An engine swap can be a useful procedure for other reasons too. Perhaps you have purchased an airplane that has a Continental 65 horsepower engine, but for one reason or another you want an 85 or a 90 horsepower engine. You can swap your 65 for an 85, or you can sell the 65 outright and purchase the higher-power plant. Actually, for restorers and homebuilders, the 65 Continental is a more valuable engine, and it is not uncommon to receive a higher price for a 65 horsepower engine than for one of higher power. This would be particularly true if the engine you are contemplating buying is of equal or higher total time since overhaul.

Some things to watch out for in engine swapping are, as stated, engines with questionable histories, engines that have been involved in accidents, engines that have a history of long disuse, storage, etc., and engines that have been used on training airplanes, particularly at higher-elevation airports—these engines will have been run hot and hard for most of their lives, and thus could mean trouble later on, even if relatively low-time.

Learning to Fly in Your Own Plane

The cost of getting into the air is getting higher with respect to getting the all-important pilot's license also. At today's prices, it can easily cost $2000 just to get a basic Private Pilot Certificate, if you are planning to use commercial Fixed Base Operator (FBO) training schools (Fig. 10-1).

Here's how the costs of learning to fly compare if one considers using a light, personally owned airplane, versus a rental airplane (all costs approximate, based on 300 hr./year):

	Yours	*Rental*
Hourly cost - airplane	$12.00	$28.00
Instructor cost/hr	15.00	15.00
Total for Aircraft/Hr	27.00	43.00

To obtain an FAA Private Pilot Certificate, you need a minimum of 40 hours, of which 20 must be solo and twenty must be dual, with an FAA certificated flight instructor. Based on this requirement, the minimum cost, based upon the above figures would be:

	Yours	*Rental*
40 hours aircraft time	$480	$1,120
20 hours dual	300	300
Total	$780	$1,420

So the cost of the *flight* training alone is almost twice as high in a modern rental aircraft as it would be in your own, and that does not take into consideration any ground training in preparation for the FAA written examination portion to qualifity for the license.

Bear in mind, of course, that in some cases an older, basic airplane might not have all the instrumentation required to complete some of the license requirements, such as radio navigation, flying by reference to instruments, etc., but that does not stop you from building time and accomplishing all the prerequisites possible in your own aircraft, and simply renting a commercial airplane to complete the remainder. Most flight schools have no problem integrating a portion of their curriculum into a course of training being pursued in a customer's own aircraft. Thus, you can see a distinct advantage in small plane ownership in obtaining your FAA pilot's license.

Resale Value

A major part of the consideration as to what type of airplane to purchase might well be the resale value of the finished, restored airplane. There are many factors that enter into the equation: type of airplane, quality of restoration, paint scheme, engine time, and even the area of the country you're in. I will attempt to give you some guideline as to what factors affect resale value and how to optimize your investment in an appreciating airplane. This translates into flying costing you less per hour, since your airplane might well be gaining value as you fly, just as stated earlier.

The type of airplane that has been discussed in these pages is essentially the two four-place fabric-covered airplane built in the 1940s and 50s. Obviously, they aren't making any more of them, and for that matter, there is nothing being manufactured that even comes close to these aircraft in terms of fun and economy. So, by the dictates of the law of supply and demand, you are dealing with a limited commodity for which the demand becomes greater as the price of fuel and new production airplanes escalates.

A Piper J-3 Cub that originally sold for $1995 brand new will bring $12,000+ today, even with 4000 airframe hours and a midtime engine. That is, as long as it is well maintained, has fairly recent fabric, and, perhaps above all, *is painted in the original "Cub Yellow" paint scheme*, complete with jet-black "lightning bolt" trim. Any other color is just not "Cub." On the other hand, a Taylorcraft BC-12D, built contemporaneously with the legendary Cub, can be painted practically any color and will still bring about $6500 to

$8000 on the market, even in outstanding condition, and moreover, even considering that the T-Craft is a faster, more economical and more comfortable airplane. An Aeronca 7AC might bring an owner $9000 - $12,000 if it is exceptional, has brand-new fabric, and a mid or low-time engine. What this means is that most older airplanes have not achieved the cult status that the original J-3 has, but they do still bring vastly higher prices that they did when new, taking into account inflation, etc.

When you get into the larger airplanes (the Stinsons, Cessna 170s, etc.), they tend to level off. They will probably not sell for a great deal more than you put into them. They are not really classics; they are not that much more economical than contemporary craft.

The quality of the restoration job you do will affect the resale value greatly. Keep this in mind as you work through the project. An additional hour here, an additional bit of sanding and doping, a bit more effort on trim or upholstery, all these things translate into dollars and cents when the time comes to sell the airplane.

If you live in the southwestern portion of the country, your airplane will bring optimum at resale, since most buyers tend to feel that the warmer, dryer climates are better for airplane preservation. By the same token, if you reside in the industrial north or northeast, your aircraft may bring less money, since the cold, air pollution, proximity to salt air, etc. will create suspicions about rust, corrosion, fabric life and a host of other maladies. This factor may be minimal, but it does exist, as does the factor of the economic realities of the geographical area. If you live in a depressed area, the market for your airplane will be limited, while those residing in wealthier parts of the country will have less trouble selling and receive more money in the end.

All these factors should be taken into consideration as you contemplate the purchase, accomplish the restoration, and anticipate the true, actual cost of owning and flying a light airplane.

Maintenance: What You Can Do

When trying to determine what you, as a non-mechanic airplane owner/restorer can legally (and safely) do with your airplane, it sometimes poses a problem, since the line between owner maintenance, work that requires a mechanics "sign-off" and that work which can be done under the direct supervision of a licensed A&P mechanic is sometimes not distinct.

Here is a guideline that covers some of the more frequent tasks that the restorer/owner may encounter, along with a listing of what

Fig. 10-1. The cost of learning to fly in one of these newly manufactured airplanes is becoming prohibitive—getting off the ground is cheaper and usually more fun in the older, more affordable planes.

can be done without the help of a mechanic, what may be done with the supervision and sign-off by an A&P, and what should really be left for a mechanic to either do, or very closely supervise and assist you to do.

	You May Do	Mechanic Must Supervise/Sign	Mechanic Should Do
1. Cleaning and polishing	X		
2. Oil change	X		
3. Oil screen check and resafety			X
4. Top overhaul and/all cylinders			X
5. Aircraft dismantling	X	X	
6. Preparing airframe for cover	X	X	
7. Inspecting airframe before cover			X
8. Sandblasting airframe	X	X	
9. Replacing wood formers	X	X	

	You May Do	Mechanic Must Supervise And Sign Logbook	Mechanic Must Do
10. Repair/weld tubing			X
11. Prime metal prior to recover	X	X	
12. Apply fabric	X	X	
13. Spray dope	X	X	
14. Sand dope and paint	X	X	
15. Modify fuselage with dorsal fins, etc.	X	X	
16. Rig controls/wings, etc.	X	X	
17. Major engine overhaul			X
18. Overhaul Propeller			X*
19. Trammel wings	X	X	
20. Splice wood spar			X
21. Route wiring/control cables	X	X	
22. Adjust control cables	X	X	
23. Change tire	X		
24. Replace instruments	X	X	
25. Repair instruments			X**
26. Cover wings	X	X	
27. Paint exterior	X		
28. Replace interior upholstery, headliner, etc.	X		
29. Repair/replace plexiglass	X		
30. Set rigging/trim			X
31. Set engine time/magnetos			X
32. Perform differential compression check	X	X	
33. Change spark plugs	X	X	
34. Clean/gap spark plugs	X	X	
35. Weigh airplane	X	X	
36. Correct weight and balance			X
37. Reassemble aircraft	X	X	
38. Test fly	X	X	
39. Sign appropriate logbooks and Form 337			X
40. Officially return aircraft to service			X***

*An FAA Approved Propeller Repair Shop must accomplish and sign prop overhaul.
**An FAA Approved Instrument Repair Station must accomplish repair/overhaul of instruments.
***An FAA Airframe Inspector must sign the Return to Service.

Fig. 10-2. Keeping your airplane hangared and covered by a protective tarpaulin, as is this obviously pampered Citabria, will pay considerable rewards in terms of appearance, long life, and high resale value. Most airplanes in this category, appreciate, rather than depreciate with age.

Summary

The dream of every pilot is to someday own an airplane. At today's prices, a new aircraft is out of reach for many sport flying enthusiasts and the cost of operating and maintaining a sparkling new bird is, for many, prohibitive. This does not mean that ownership of a sport airplane is forever the privilege of the wealthy.

By looking to the older used airplanes, you immediately discover several encouraging facts: (1) older airplanes are, in many ways, more fun to fly; (2) there are many older fabric-covered aircraft on the market; (3) these aircraft, when compared with current production models, are less costly to operate; and (4) probably the most attractive aspect of this alternative to new aircraft ownership is that they are available at a fraction of the cost of a new airplane.

Specifically, the small airplane that is in need of restoration can often be purchased for $4000 or less!

There are many pitfalls, however, to be aware of when considering this method of acquiring an airplane and it is not by any means an easy way to do so. But armed with ample information and knowledge of the rules, regulations, requirements, procedures and costs, restoring a light airplane can be a fun and cost-effective way to get airborne. Happy flying!

Appendix
State Aviation Departments

Alabama Department of Aeronautics
Room 627, State Highway Building
11 South Union Street
Montgomery, AL 36130

Alaska Department of Transportation and Public Facilities
Pouch 6900
Anchorage, AK 99502

Arizona Department of Transportation
Aeronautics Division
205 South 17th Avenue
Phoenix, AZ 85007

Arkansas Division of Aeronautics
Adams Field
Old Terminal Building
Little Rock, AR 72202

California Transportation Commission
Aeronautics Subcommittee
1120 North Street
Sacramento, CA 95814

Colorado State Patrol Aircraft
4201 East Arkansas Avenue
Denver, CO 80220

Connecticut Department of Transportation
Bureau of Aeronautics
24 Wolcott Hill Road
Wethersfield, CT 06109

Delaware Transportation Authority
Aeronautics Section
P.O. Box 778
Dover, DE 19901

Florida Department of Transportation
Aviation Bureau
Division of Public Transportation Operations
605 Suwannee Street
Tallahassee, FL 32301

Georgia Department of Transportation
Bureau of Aeronautics
5025 New Peachtree Road, N. E.
Chamblee, GA 30341

Hawaii State Department of Transportation
869 Punchbowl Street
Honolulu, HI 98613

Idaho Division of Aeronautics and
Public Transportation
3483 Rickenbacker Street
Boise, ID 83705

Illinois Division of Aeronautics
Capital Airport
Springfield, IL 62706

Indiana Aeronautics Commission
Suite 801, State Office Building
100 North Senate Avenue
Indianapolis, IN 46204

Iowa Department of Transportation
Aeronautics Division
State House
Des Moines, IA 50319

Kansas Department of Transportation
State Office Building
Topeka, KS 66612

Kentucky Department of Transportation
Division of Aeronautics and Airport Zoning
419 Ann Street
Frankfort, KY 40601

Louisiana Department of Transportation and
Development, Office of Aviation and
Public Transportation
P.O. Box 44245 Capitol Station
Baton Rouge, LA 70804

Maine Department of Transportation
Bureau of Aeronautics
Transportation Building
Child Street
Augusta, ME 04333

Maryland Department of Transportation
State Aviation Administration
P.O. Box 8766
Baltimore/Washington Intl. Airport, MD 21240

Massachusetts Aeronautics Commission
Boston/Logan Airport
East Boston, MA 02128

Michigan Aeronautics Commission
Department of Transportation
Capital City Airport
Lansing, MI 48906

Minnesota Department of Transportation
Aeronautics Division
Room 417 Transportation Building
St. Paul, MN 55155

Mississippi Aeronautics Commission
500 Robert E. Lee Building
P.O. Box 5
Jackson, MS 39205

Missouri Highway and Transportation Department
6th Floor, Broadway Building
P.O. Box 1250
Jefferson City, MO 65101

Montana Board of Aeronautics
P.O. Box 5178
Helena, MT 59601

Nebraska Department of Aeronautics
Municipal Airport
P.O. Box 82088
Lincoln, NE 68501

Nevada Public Service Commission
Kinkead Building
505 East King Street
Carson City, NV 89710

New Hampshire Aeronautics Commission
Municipal Airport
Concord, NH 03301

New Jersey Department of Transportation
Division of Aeronautics
1035 Parkway Avenue
Trenton, NJ 08625

New Mexico Department of Transportation
Aviation Division
P.O. Box 579
Santa Fe, NM 87503

New York State Department of Transportation
Airport Development Section
1220 Washington Avenue
Albany, NY 12232

North Carolina Department of Transportation
Division of Aviation
P.O. Box 25201
Raleigh, NC 27611

North Dakota Aeronautics Commission
Municipal Airport
P.O. Box U
Bismark, ND 58505

Ohio Department of Transportation
Division of Aviation
University Airport
2829 Granville Road
Worthington, OH 43085

Oklahoma Aeronautics Commission
424 United Founders Tower
Oklahoma City, OK 73112

Oregon State Department of Transportation
Aeronautics Division
3040 25th Street S.E.
Salem, OR 97310

Pennsylvania Department of Transportation
Bureau of Aviation
Harrisburg International Airport
45 Luke Drive
Middletown, PA 17057

Rhode Island Department of Transportation
Division of Airports
T.F. Green State Airport
Warwick, RI 02886

South Carolina Aeronautics Commission
Drawer 1987
Columbia, SC 29202

South Dakota Department of Transportation
Division of Aeronautics
Pierre, SD 57501

Tennessee Department of Transportation
Bureau of Aeronautics
P.O. Box 17326
Nashville, TN 37217

Texas Aeronautics Commission
40 East 5th Street
Austin, TX 78701

Utah Department of Transportation
Division of Aeronautical Operations
135 North 2400 W.
Salt Lake City, UT 84116

Vermont Agency of Transportation
State Administration Building
133 State Street
Montpelier, VT 05602

Virginia Department of Aviation
4508 South Laburnum Avenue
P.O. Box 7716
Richmond, VA 23231

Washington State Department of Transportation
Division of Aeronautics
Boeing Field
8600 Perimeter Road
Seattle, WA 98108

West Virginia State Aeronautics Commission
Kanawha Airport
Charleston, WV 25311

Wisconsin Department of Transportation
Division of Transportation Assistance
Bureau of Aeronautics
P.O. Box 7914
Madison, WI 53707

Wyoming Aeronautics Commission
State of Wyoming
Cheyenne, WY 82002

Sources of Materials and Supplies

The Aeroplane Store
Kampel Enterprises, Inc.
Wellsville, PA 17365
Telephone: (717) 432-9688
Offers a wide range of supplies for the rebuilder and homebuilder.

Aircraft Components, Inc.
700 North Shore Drive
Benton Harbor, MI 49022
Telephone: (800) 253-0800
(616) 925-8861 (in Michigan, Alaska, Hawaii)
Suppliers of a wide variety of aircraft and pilot supplies.

Aircraft Supply
Allegheny County Airport
West Mifflin, PA 15122
Telephone: (800) 245-0690
Large supplier of aircraft parts and components, aircraft fabrics and finishes.

Airtex Products, Inc.
259 Lower Morrisville Road
Fallsington, PA 19054
Telephone: (215) 295-4115
Supplies a line of interiors and textile products for airplanes.

All Aircraft Parts
16673 Roscoe Boulevard
Sepulveda, CA 91343
Telephone: (213) 894-9115
Supplies a large variety of aircraft parts. They can also supply custom-made items.

Ceconite, Inc.
4677 Worth Street
Los Angeles, CA 90063
Telephone: (213) 269-1475

A supplier of a patented Ceconite aircraft recovery process and synthetic fabric.

Cooper Aviation Supply Company
2149 East Pratt Boulevard
Elk Grove Village, IL 60007
Telephone: (312) 364-2000

A supplier of aviation products, specializing in recovery materials and supplies.

El Reno Aviation, Inc.
1004 S. Country Club Road
P.O. Box 760
El Reno, OK 73036
Telephone: (405) 262-2387

A supplier of engine components and miscellaneous hardware.

Fresno Airparts Company
Chandler Field
520 West Kearney
Fresno, CA 93706
Telephone: (209) 237-4863

Source of engine parts, engine instruments, tires, fabric supplies.

Hower Aviation
4822 Royal Palm Avenue
Sarasota, FL 33580
Telephone: (813) 355-5237

Suppliers of aircraft recovering materials, etc.

Randolph Products Company
P.O. Box 67
Carlstadt, NJ 07072
Telephone: (201) 438-3700

Suppliers and manufacturers of aircraft dopes and finishes.

Stits Aircraft Coatings
43rd and Fort Drive
P.O. Box 3084
Riverside, CA 92509
Telephone: (714) 684-4280

Manufactures and supplies a synthetic recovery process.

Univair Aircraft Corporation
Route 3, Box 59
Aurora, CO 80011
Telephone: (303) 364-7661

Specializes in the manufacturing and distributing of aircraft parts, supplies and materials. They also publish manuals and parts books.

Wag Aero, Inc.
Box 181, North Road
Lyons, WI 53148
Telephone: (414) 763-9586

Manufactures and distributes numerous aircraft parts and accessories. They also offer many service manuals and parts lists for out-of-production aircraft.

Wicks Aircraft Supply
410 Pine Street
Highland, IL 62249
Telephone: (618) 654-7447

Offers aircraft wood, tubing, fiberglass cloth, resins and glues.

Index

Index

A

Adel clamps	104
Aeronca Champion	8, 11, 14, 20, 25
Aircoupe	23-24, 27
Aircraft Owners and Pilot's Association (AOPA)	2
Airframe inspection	66-98
Airport charges	3, 10
Airworthiness Certificate	36, 42
Airworthiness Directives	6
Aluminum, painting	138-140
Aluminum surfaces, new	138
Antique aircraft	48

B

Bill of sale	3, 42
Biplanes	48
Bolts, aircraft	106

C

Caveat emptor	48
Ceconite	8, 109-134
Cessna	11, 20, 28, 30
Chromated enamels	138
Cock pit inspection	37
Color application hints	133
Compression	33
Continental engines	31, 32, 35
Control cables	107
Control surface CG	113
Control surface preparation	98
Conventional airplane alternatives	154
Cooper Aviation Supply Company	8
Corrosion	72-91
Corrosion protection	143
Costs	
inspection	6
insurance	4
initial	vii, 1
maintenance	4-7
operating	2-4
per hour	8-10
rebuilding	7, 8
sport planes	11
tie down	3, 6

D

Damaged aircraft	160
Differential compression	33
Disassembly	54-66
Distortion, airframe	122
Dye-Check	97

E

Electronics	149-154
ELT	154
Engine inspection	36
Engine logs	32-34
Engine overhaul	32, 35
Engine swapping	161-162
Ercoupe	23-24, 26, 27
Exchange price, engine	161

F

FAA approval	109-111
Fabric attachment spacing	117
Fabric-covered aircraft	
availability of	11
choice of fabric for	50
inspection of	7
Fabric finishing,	
Ceconite 7600	130, 134
Dac-Proof and Spra-Fill	129
standard	126
Fabric, pre-sewn	109
attaching methods	114-115
Fabric removal	62
Fiberglass, painting	144
Filler application hints	133
Finishing coats	144
Flight test	36, 40-42
Flight training	162-163
Floatplanes	147
Franklin engines	31
Fuel costs	2
Fuselage cover	104-108
Fuselage inspection	39

G

Glaze application hints	133
Grade A cotton	50-52
Grommet inspection	37

H

Hangering	2-4
Heat, shop	53
Heat shrinking, fabric	121-124

I

Inspection	6
annual	124
fabric-covered aircraft	7
Insurance	4
Interior surfaces, painting	143-144
Irish linen	50-52

K

Knots	118-122

L

Liens	2
Luscombe Silvaire	21-23, 24
Lycoming engines	31, 35

M

Magnesium, painting	141-142
Maintenance	164-166
owner	164-166
preventive	6
Maintenance checklist	166
Maintenace costs	4
Major overhaul	34-35
Metal aircraft	29
Metal finishing	137-146

N

New Airplane costs	11

O

Oil filter	34
Overhaul, major	34-35
top	34
Ownership, reasons for	vii
Oxidized surfaces	138, 142

P

Painted surfaced, old	138
Painting tips	146
Periodicals advertising used aircraft	43
Piper Colt	24-28
Piper J-3 Cub	19-20, 22
Piper PA-22	20, 21, 24, 28, 29
Piper Tri-Pacer	19-20
Plywood, covering	123-124
Porterfield	15, 16, 20
Primers	144-145
Pulley wear	103
Punch test	7, 39

R

Radios	149-154
Rare airplanes	49
Reassembly	135-136
Recovery costs	9
Registration	5, 42
Registration fee	1
Remanufactured engine	161
Repairs	124
Resale value	163-164
Rust	72-91

S

Sales tax	1
Salvage, rebuilding from	159-160
Sandblasting	90-97
Service manual	55
Skis	148
Spraying equipment	145
Steel, painting	140-141
Stinson	11, 13-18, 19
Synthetic fabrics	50

T

Tail inspection	40
Taylorcraft	12-13, 16, 17, 18, 20
Tiedown costs	3, 6
Title search	1-2
Top overhaul	34
Trade-A-Plane	43, 147, 160
Turnbuckles	105

U

Ultralights	154-158

V

VHF radios	149-154

W

Weedhopper	155-156
Wing inspection	37
Wing preparation	98
Wing removal	57-62
Wing trammelling	101-102
Wood kits	104
Work area	52-54

Z

Zinc chromate primer	98